VERTIGO!

WHEN THE WORLD SPINS
OUT OF CONTROL

Menière's Disease, Benign Positional Vertigo
And Other Vestibular Illnesses

Causes, Symptoms, Treatments, Mysteries
Remarkable & Varying Experiences With Vertigo

Linda Howard Zonana

Cover Design by Marian H Willmott

outskirtspress
DENVER, COLORADO

Dedication

This book is for the millions of people who struggle with the dreadful sensation of vertigo and its attendant symptoms.

Since I think of a dedication as an expression of love, the book is also dedicated to my husband, and my children and their spouses: To Howard, Jeremy and Sonja, Elisa and Mark, Jessica and Ariela. I also dedicate it to my sisters and their husbands: Joanie and Jovelino, Marian and Denny, and in memory of Susie. And to my grandchildren: Noa, Isaac, and Jude, with the hope that by the time any of them are old enough to read it, much more will be known about vestibular illnesses and the book will be outdated.

Note

This is a book written by a layman. It is based on considerable research, and on conversations with numerous people suffering with vertigo as well as with several professionals in the field. The chapters describing standard medical and anatomical information have been reviewed by an otolaryngologist, Victor E. Calcaterra, MD. The fact that lesser known or less well accepted ideas and treatments are discussed within the book does not indicate that he supports them. Opinions scattered throughout the work, both overt and implied, are mine and not his. Whatever mistakes or misinformation that may be found in the text are my responsibility entirely.

The book should not be viewed as a guide to the treatment of vestibular disorders, but rather as a source of information as to what some of the possibilities are. The reader should consult with a medical or health professional concerning decisions about care.

Contents

Introduction and Overview

Seven years ago I had my first attack of vertigo, and was diagnosed with Menière's Disease. I found the condition to be deeply distressing and confusing, and could find little to read about it that was really helpful to me. Most of it seemed oversimplified and left too many questions unanswered. So I decided to write a book of the sort I would like to have read at that time, one that I hope will also be of interest to families and friends of people afflicted with vertigo. Perhaps it will even have some appeal to those in the general public who are curious about vertigo, or about issues pertaining to mysterious illnesses.

Vertigo is a significant public health issue, but not a widely publicized one. About two to three percent[1] of the population will seek medical help for it at some point in their lives. "Vertigo" in itself is not a diagnosis but rather a symptom, just as a headache or stomachache might be. Sometimes the underlying cause of that vertigo will be unclear, other times it will seem to fit well in a diagnostic category. Sometimes it is caused by injuries to the head, or something going wrong in the brain, and sometimes it is caused by disturbances in the inner ear. No matter the cause, it is an enormously unsettling experience. The word "dizzy" is often used to describe it, but vertigo - especially intense vertigo - is not the same thing as dizziness.

---

1   This is not an exact percentage. Different percentages will be found in varying articles.

If the vertigo arises from the inner ear, what will take place over the next days, weeks, months, or even years may not be predictable. If it is caused by an acute infection, it may happen once and never again. But when the vertigo is thought to be a symptom of two specific inner ear disorders, the symptoms are apt to come back again from time to time - a disturbing thought, and not one that is easy to get used to. This is sometimes true of a related disorder, Vestibular Neuritis, as well, although VN is not as likely to recur. Vertigo can be difficult to treat, although some kinds respond well to current medical approaches. All too often, the first doctor a patient encounters is not very familiar with it, so many people frequent emergency rooms or return several times to private doctors before getting an understanding of what is happening to them and what they can do about it. In the meantime, they are per-plexed and worried. If they do begin a course of treatment that seems to be helping, it can be difficult to tell whether it's thanks to the treat-ment that they are feeling better, or whether this would have happened anyway, because of the way vertigo tends to come and go.

I received my diagnosis of Menière's Disease, which arises in the inner ear, early on. I understood why I was given the diagnosis – I could see that I had the requisite symptoms – but nonetheless I didn't feel as if the doctors actually understood what was happening. Much of what I read about it on various Internet sites did not seem to apply to me. I found nothing that fully described what I was actually going through. The web pages, articles, and books geared toward the public seemed cut-and-dried, and didn't address my questions adequately. Neither did pamphlets available as handouts in doctor's offices. The processes in your inner ear and brain that lead to vertigo are not yet fully understood, but a lot of what I read described them without ac-knowledging their complexity or the gaps in knowledge. In addition, there was little recognition of the tremendous challenge confronting people with vertigo, especially those who suffer repeated bouts of it.

At the time vertigo began in my life, I was in my 60s and working as a social worker in an elementary school, a hectic and interesting job I had had for many years. I had no interest in retiring – I loved working with the children and their parents, as well as with the staff. I enjoyed the variety of problems that were brought to my attention, ranging from a child upset because a pet had died or children having trouble getting along together, to children who had more long-standing sorrows or behavior problems, some of these serious. My husband, a psychiatrist, was deeply engaged in his own work. Our household consisted of just the two of us, since our three children were out living on their own, with their own jobs, partners, and homes. We were both in fine health and able to enjoy energetic activities such as hiking and skiing, as well as plenty of quieter ones. At first, it seemed as if Menière's Disease posed a major threat to all that, and to my life as I knew it.

I wanted to learn more about it, in order to write a book that would provide a detailed account of what is known about the causes, treatment, and course of the diseases that include vertigo. I wanted to present a comprehensive picture of what it is actually like to live with vertigo. I wanted to find out how other people experienced it and how it affected their lives. Thus, I interviewed more than 50 people who are afflicted with vertigo, to learn about their exact symptoms, their experiences in trying to find help, what has or has not been helpful, and the impact of vertigo on their lives. These were mostly phone interviews, so I was able to talk with people all over the country. In this way, I was able to obtain a perspective that transcended the usual clinical descriptions of these conditions.

I found a surprising number of subjects through the web of personal acquaintance (a friend or relative of a friend). Secondly, I put out a notice through alumnae magazines of schools I had attended, asking for people with vertigo to contact me if they were willing to be interviewed. This was my largest source of subjects. I was also able to place a notice on the Vestibular Disorders Association (VEDA) website. I talked with people of all ages, who were affected in varying degrees by vertigo and its accompanying symptoms. I was struck with how common it was for people to struggle in order to find support and help, and how long some had to wait for a diagnosis or any sort of understanding concerning what was happening to them, how frustrating many had found their contacts with doctors to be. Overall, most perceived a lack of urgency from the medical profession, and felt strongly the need for more research into effective therapy.

My aim in this book is to concentrate on the two of the most commonly diagnosed conditions that originate in the inner ear, or the "**vestibular system**," where the organs that help maintain balance are situated. These conditions are known as "Menière's Disease" and "Benign Paroxysmal Positional Vertigo," or BPPV, a variant of this being "Benign Positional Vertigo," or BPV.[2] More briefly, I will also explore two diseases characterized by vertigo that seem to be caused by an acute infection in the area, usually viral. These are vestibular neuritis, or VN, and labyrinthitis. Because some of my subjects suffered from vertigo secondary to head injuries, a chapter will be devoted to experiences people have had with brain-based vertigo. In addition to providing a description of the current understanding of these conditions and the most commonly accepted ways to treat them, I will include many real life accounts to illuminate the varieties of experience people have. These are scattered throughout the text, along with a memoir of my own experiences with vertigo. The latter are located in separate chapters, for the most part. In addition, I will explore some of the more controversial ideas about cause and treatment.

It is important to remember that the basic cause of vertigo is not always found in the inner ear. People with conditions such as brain tumors, stroke, high blood pressure, or acoustic neuroma, among many others, may develop vertigo. Blows to the head may cause vertigo, sometimes long after the injury is over. Symptoms arising within the brain are not necessarily more ominous than those arising in the inner ear, despite how frightening it sounds. A good number of people who are subject to migraine headaches also suffer with a related vertigo, called migraine associated vertigo (MAV.) Figuring these things out and providing effective treatment can be a challenge. Thus, anyone

---

2   Throughout this book, I will refer to these conditions by initials: MD for Menière's Disease, and BPPV for Benign Positional Paroxysmal Vertigo. VN will be used for Vestibular Neuritis.

who has an acute episode of vertigo or dizziness or repeated milder ones, should see a physician in order to establish a diagnosis, as well as to assess what can be done to alleviate the symptoms.

I prepared for writing this book in several ways. I did extensive research online, read several books, and had access to many professional articles through MedLine. In my interviews, I did not restrict my subjects to specific diagnoses. It turned out that about one third of the subjects had Menière's, one third some version of BPPV, and the remaining third varied: some suffered with vertigo but did not have a diagnosis, some had had an injury to the brain, some were diagnosed with vestibular neuritis. I also interviewed several professionals in the field.

Among those I interviewed, about a quarter were men. Mine was probably a skewed population, since according to other sources, the conditions tend to affect both sexes about equally. In my sample, the age of onset seemed random, sometimes beginning in a person's twenties, right on up to seventy. Among those with whom I spoke, the majority had reached the age of at least fifty, however. The youngest person I interviewed was 30 at the time. Again according to other sources, these conditions usually start in one's 30s and 40s. I developed a series of questions to help structure the interview, which was arranged as an appointment, and spent anywhere from half an hour to two hours on the phone with my subjects. Most commonly the interviews lasted around an hour.

Some of the information presented in this book may seem discouraging, not only to people suffering with vertigo but to those who care about them. Remitting diseases, which is to say ones that come and go, are unsettling to everyone. It is important to remember that while vertigo is indeed is a capricious and unpredictable condition, it comes in many forms. Thus it is very unlikely that any two people will have exactly the same experience with it. There are those who have one or two major episodes, then no more. There are those who have symptoms for a while which then fade, or which are successfully addressed by some kind of treatment. There are those who are subject to repeated attacks, but at widely spaced intervals, who feel absolutely fine the rest of the time. There are those whose symptoms are not overwhelming to them. The people who are the most challenged by the disease are those who have intense vertigo often, or who feel unwell most or all of the time, so that in essence they suffer from a chronic illness. How the condition in a given person will play out over time cannot be known for sure at the outset. In doing research for this book, it turned out that the majority of people I interviewed did have intermittent attacks over extended periods of time, often years. Even so, most of them feel better now than they did at first, which has been true for me as well. It is a burden for anyone to think of dealing with vertigo down over time, but I have found it helpful to learn more about it, and hope others will also find this to be true. I continue to feel hopeful about possible treatments that may become available.

The book will provide comprehensive, detailed information written in a readable style. I will not be delving into every idea that may be found on the Internet or the literature concerning the variety of diseases that may include vertigo as a symptom. There will not be an exhaustive discussion of every remedy some may believe has therapeutic value, although I will review most of those that people in my study found helpful. My overall effort will be to provide an understanding of vertigo itself, of current medical thinking about these conditions, and the most prevalent or common therapeutic approaches. In addition I will explore some of the less conventional treatments, as well as the psychological and emotional effects of dealing with an illness that tends to recur. The final chapter is a brief history of the development of mankind's understanding of vertigo and how to treat it, and how it relates to the workings of the inner ear.

Work on the book was begun in 2010. Most of the interviews were conducted in 2011 and 2012. The writing was completed in 2013. Names and other identifying information about most participants in the study have been changed. With their permission, I did not disguise the identities of two of the subjects, since there were particular reasons to include that information in describing their experiences.

Please note there is a glossary at the back of the book that gives brief definitions of technical terminology found in the text.

1

Vertigo Strikes! My First Experience

Monday, February 20, 2006: Home on winter break from my job as a social worker in a school, I was down in the basement stuffing clothing into bags to take to Goodwill, when I experienced a sudden wave of dizziness. Afraid that I was about to faint, I sat down on the floor. The feeling intensified, so I lay face down, assuming it would pass soon. But every time I moved my head, or even my arm, I was overcome by a vigorous churning within my head, as if my brains were in a clothes washer, sloshing around and around. My eyes lost focus and I felt removed from my own self. I kept my eyes lowered, looking only at the ugly worn carpet beneath me. Curiously, I wasn't very frightened, guessing that I might have an ear infection, since I'd had a stuffy head all weekend. I wondered whether to stay where I was, possibly hours until my husband returned home, or whether to try to get upstairs to the phone. I was scarcely able to sit and certainly unable to stand, and first considered crawling out through the garage and around the house to the back door. I imagined the possibility of collapsing, unable to go on, and being stuck outside in the gray frozen day. I was worried about falling on the stairs, too, but decided to try creeping toward them. Since they have open risers, I was able to cling to each step and drag my legs up after me. At the top I experienced a burst of nausea and vomited. I crawled to a chair near a telephone and, sitting with my head in my hands, unable to open my eyes without increased nausea, I vomited repeatedly over the next two hours as I waited for my husband to return my call. I remember saying to him, "I'm completely non-functional." He came home immediately to take me to the "Urgent Visit" at our health plan, and held me up as I attempted to walk. Even then I was so unsteady that I frequently knocked him off balance.

Being interviewed by the doctor was socially very uncomfortable. I was not able to look at him, since I couldn't fix my gaze on anything, nor could I smile or talk in a normal manner. My voice sounded like a flat drone in my ears. He spoke with my husband about wanting to rule out the possibility that I was having a stroke, which surprised me, since I didn't think I had symptoms of one. I knew that when people are dizzy their eyes dart uncontrollably back and forth, a condition called "nystagmus" and asked the doctor if I had that, which he confirmed. He also confirmed that I had a disturbance in my inner ear, which might be due to an acute viral infection: "labyrinthitis." During the time I was there, the dizziness began to ease a bit, and was pretty much over with by the time we got home. The whole horrible episode lasted about five hours, but at the time my hope, and probably my assumption, was that it would fade into memory, like a bad flu or a twisted ankle.

I was exhausted the next day and had little ambition. Then, in the evening I experienced a much milder dizziness. Throughout the following week I felt mildly unwell most of the time, but able to do most normal activities. Nearly every day at random times, a dizzy sensation would float into my head, gaining force over a few seconds so that my brains again felt as if they were churning. Once it got going, it lasted an hour or two. It was sickening and disconcerting, but nowhere near as dramatic as the first time. I was reminded of those characters in comic books who have wavy lines and circles drawn over their heads to indicate they are reeling from a punch. Those wavy lines seemed to swirl inside my own head.

<u>Monday, February 27, 2006</u>: School, and hence my job, resumed after winter break. I was afraid to drive, and so made plans with my

husband and friends to cart me back and forth. I warned people at work that I might have to hold on to the walls from time to time in order to get around. In the meantime, we had also made an appointment with an ENT doctor (ear, nose, and throat specialist) for late in the afternoon of that day, to evaluate my lingering symptoms. He shocked me by asking whether I'd had a big pepperoni pizza the night before the first attack. I had no idea why he would ask such a question, and thought that if a large pizza were the culprit, half the country would be lying on the floor on any given day, throwing up as the room whirled around. A little later, he explained that excess salt in the diet is thought to contribute to vertigo in people with Menière's Disease, which is what he suspected I had. During the course of the week that followed, I had several appointments to determine whether that was my diagnosis. I had a hearing test, a test of my balance, an electroencephalogram (EEG). Later in the spring, I also had an MRI to make sure my symptoms weren't caused by a tumor. It turned out that I had a 37% hearing loss in my left ear, something consistent with Menière's, which the other tests seem to confirm as well. I had heard a little bit about Menière's Disease before, and because it is a recurring condition I received this news with a sinking heart, as well as some level of disbelief. During these few days since the onset of symptoms, a couple of people had told me of relatives who had been in bed for a few hours with vertigo which then went away, never to return. This, of course, was my preference for myself.

At that first visit, a brief course of oral steroids was prescribed for me, to be taken over a six day period. Although I did gradually improve over the course of the next couple of weeks, there was no dramatic change (such as you often see in conditions that respond to antibiotics). In addition to the steroids, a low salt diet and a diuretic[3] were recommended, but despite these I continued to have various troubling experiences for weeks. I frequently stumbled around like a drunken sailor, sometimes losing my balance when I wasn't anticipating it. It was nearly impossible to walk in the dark without staggering. Sitting

---

3   Details about treatment recommendations can be found in Chapter 7.

down fast might precipitate an intense (although temporary) whoosh of dizziness. It was a good idea to lie back carefully after getting in bed to avoid another such whoosh. There were times when I experienced rather loud tinnitus (noise inside your ear or head.) The tinnitus was most pronounced about a month after this all began, during a time when I was feeling dizzy and miserable a good part of the day. Most of the time it was a high steady note I could mostly ignore, but in bed in the quiet of the night, the sound felt invasive. During the day, when I wasn't actually dizzy my head tended to feel funny, as if it were the middle of the night and I'd had too much to drink. And beyond all of this, I felt extraordinarily tired a great deal of the time. I worried about whether I could keep my job if this continued.

Thus began a search that has continued for the past several years – a search to cope with and to understand just what is happening, and a search to find out what can be done about it. Working on this book is part of that search.

2

Vertigo and Dizziness

Vertigo and dizziness are not completely interchangeable terms - there's a fine line between them, although vertigo is a form of dizziness. Everyone has had some experience with feeling dizzy. It is common when you stand up suddenly to feel light-headed and mildly off balance for a few seconds. You may feel it when you are sick, low on sleep, very tired, or when you've gone without food or drink for too long. Some people can't stand near the edge of a deep ravine, or even think about standing near one, without feeling faint and weak-kneed. Being anxious often causes it. Sometimes it's even fun, such as when you experience it on an amusement park ride, or when children twirl around and fall on the grass just to get the sensation. Dizziness may make a person feel faint or spaced out, and when it is strong may be accompanied by poor balance and a feeling of nausea. Most of the time, if you are mildly dizzy, it doesn't last very long.

Although everybody thinks they know what "dizzy" means, it's actually a challenge to find words to describe it. "Vertigo" may be somewhat easier that way. For one thing, vertigo is more active – characterized by a strong sensation of movement. One formal definition is that vertigo is "the perception of motion when there is none." Many people say they feel as if the room is spinning around – this is the most common description. It happens to many people when they've had too much to drink. There are degrees of vertigo – mild vertigo will impair balance, but you'll still be able to stagger around on your own. (Think of the police officer checking to see whether the erratic driver is able to walk a straight line.) Intense vertigo will make it impossible for you to maintain any balance at all, such that you will be unable to walk without support, and possibly not even then. When the world is experienced as "spinning," it will look as if it is rotating, and the person will feel better if s/he closes his or her eyes. But closing one's eyes does not usually eliminate the sensation – the feeling of whirling around will continue, but be slightly less sickening.

Reading about vertigo sometimes sounds cut and dried, as if the statement that you feel as if the room is spinning around gives a full picture of what it's like. But even in its most dramatic forms, it is not experienced in just the same way by everybody. Sometimes whirling or spinning aren't exactly what's happening. In my case, rather than feeling the room rotating around me, vertigo was primarily experienced as occurring *inside* my head. It seemed that my brains were churning around in a random rather than a rotary pattern. In looking over my notes on my interviews, there were several others who also felt the vertigo inside their heads, although this was less common. I don't remember any of the doctors I consulted asking for details about the vertigo, but I do wonder if the differences could indicate something different about the underlying mechanisms, and possibly about a diagnosis.

Vertigo is very nauseating, regardless of how it's experienced, and people are apt to vomit repeatedly during an attack that lasts a while. Some lie on the floor in the bathroom, waiting for the next wave of nausea to come, so they can heave themselves to their knees and use the toilet. Others may keep plastic bags or wastebaskets nearby. Your autonomic nervous system becomes active, stirring up some of the "fight or flight" responses. Thus people may experience an increased heart rate, faster breathing, sweating (known, for some reason, by the fancy term **diaphoresis**) - all those functions that your body marshals when you are in danger. For some people, the gastro-intestinal system becomes overactive, so they get diarrhea in addition to the vomiting. Thinking or concentrating on anything besides all the horrible things that are going on is just about impossible.

The duration of a vertigo attack varies enormously. It is common for them to last for a few hours, but sometimes it is a day or two before they simmer down. A few people reported even longer episodes to me. It is also common for people to experience sudden brief attacks, perhaps lasting no more than a minute. Sometimes these will occur in clusters, meaning you might have several in a day, or have a few days when you will be prone to them, before they vanish. These were dubbed by one person as "mini-whirls." Many people never experience any long episodes, but probably most who are prone to extended bouts of vertigo are likely to have occasional short ones as well.

When people have vertigo, their eyes dart back and forth uncontrollably, a phenomenon called **nystagmus**. I first learned the word from my father when I was a teenager. Our cat was lying on his side on the linoleum floor of the kitchen, and my father spun him around a few times (without picking him up). When the cat stood up, he staggered for a few steps, and my father told me to notice the jerky motion of its eyes. Not a very nice thing for him to have done, but the cat was back to normal in moments, and did not seem traumatized. Thus, when I had my first bout of vertigo, I was pretty sure I had nystagmus, as the doctor confirmed. Probably because of nystagmus, people in the throes of vertigo prefer to look down or keep their eyes at least partly closed. The world looks jumbled and disturbing to a person who can't focus on anything because their eyes are in constant motion.

Although you will often see vertigo referred to as a "balance disorder," during an attack most people are mainly aware of how dreadful they feel inside their heads. Even a momentary surge of vertigo, maybe from moving your head the wrong way or sitting down too suddenly, is enough to be strongly distressing, similar to what you might experience with a jolt of sharp unexpected pain. All vertigo, whether long or short in duration, mild or strong in intensity, is disconcerting and incapacitating at some level. As noted above, during a strong attack a person will be unable to walk, and possibly not even sit up, unassisted.

Then there are people whose vertigo lasts for hours, but their equilibrium is less affected – they are able to walk around as long as they can hold onto the wall or furniture. Others may feel woozy and off balance, but not so severely they can't get across the room. However, in so doing they may need to pay close attention to every move they make, and feel sick and unable to think clearly as long as it lasts. Like anything else, the intensity of symptoms for any given person is likely to vary from time to time, and for many, the worst attacks are the first attacks.

It is worth noting here what happens in the aftermath of an attack. Unless the attack is the brief kind, people are likely to feel exhausted afterward and need to sleep. In many cases they will feel normal after the nap or by the following day. But for others the attack has a lingering effect. They may continue to feel unwell, vaguely nauseated, fatigued. Their balance may be shaky. They may be subject to brief sensations associated with vertigo, such as funny noises, zings, or pressure inside the ear or head. They may find it hard to concentrate during such times, or find dealing with myriad demands extra stressful. This state of affairs can drag on for longer than you would like before fading away. (In my initial investigations I was not able to find a statement anywhere about symptoms continuing for days or weeks after a vertigo attack.)

Some people are given the impression that vertigo is a diagnosis, but as noted earlier, it is a symptom of several possible illnesses or conditions. It is frequently a major challenge to figure out which one is the culprit. In later chapters, I will present more detail about the experience of vertigo as well as of imbalance.

3

Diagnosing Vestibular Disorders

Some people have a vertigo attack and decide to wait and see what happens before seeking medical help. But unless the attack is brief and fairly mild, most are worried and often frightened, and contact a physician or visit an emergency room as soon as they can. Whether this initial visit leads to an explanation as to what is wrong, a diagnosis, and recommendations for treatment varies considerably. Sometimes a referral is made to an ear, nose, and throat specialist (an ENT[4] or **otolaryngologist**) for a more thorough evaluation before a conclusion is reached. But all too often, things remain unclear to the patient, and in many cases, presumably to the doctor as well. In this chapter, we will go over some of the criteria for settling on a diagnosis when the symptoms seem to arise from the inner ear. There is an overlapping of symptoms among all these disorders, so that it is not always easy to be certain about the diagnosis.

The part of the inner ear that contains the semicircular canals,[5] whose function it is to regulate balance, is called the **vestibular system**, so that things that go wrong within the system are called vestibular disorders. The vestibular system is located within the temporal bone

---

4   An **otologist** is an ENT who specializes in the ear itself and would be the subspecialty most familiar with inner ear conditions. A **laryngologist**, also an ENT, specializes in the throat.

5   These are three loops, set at different angles, that lead down to the nerve.

at the base of the skull, making it very difficult to study directly. It shares space and interacts with the auditory system, since it directly adjoins it. Four main conditions have been identified as disorders originating within this system. They are benign paroxysmal positional vertigo (BPPV), Menière's Disease (MD, the diagnosis I was given,)[6] labyrinthitis, and vestibular neuritis (VN.) Labyrinthitis and vestibular neuritis are both infections, usually viral in origin. There is speculation that BPPV and MD may also develop as a result of a virus, but there is not general agreement concerning this. As you search through the Internet, there are fewer "hits" and far less information about VN and labyrinthitis available than there is for BPPV or MD. Often, they are lumped together under one title and don't even rate a separate article. Possibly this may have to do with the fact that they are both understood as infections, so that cause and treatment may seem clearer.

It is important to note that, in all four conditions, typically – and luckily - only one ear is affected. The sensation of vertigo is not experienced as being one-sided, however. It is said that about 20% of people with MD[7] will develop involvement in both ears, although this may not happen until years after the onset of the disease. It's a disturbing thought, but at least the odds are in one's favor.

In thinking about the diagnoses associated with vertigo, what is believed to cause them, and how they work, it is important to bear in mind a few things. One is that, when you read about MD or BPPV in pamphlets, articles, books, or on the Internet, the lists of symptoms provided tend to make it sound as if reaching a conclusion about diagnosis is clear-cut and simple. Another is that you get the impression that people who suffer from vertigo have pretty much the same experiences. Far from being helpful to me, I found such descriptions of these conditions disconcerting, since they didn't fit with my own experience. Many of the people with whom I spoke felt the same way. Most articles about MD state that people generally have an attack lasting a

---

6   Remember the initials MD and BPPV will be used throughout the text for Menière's Disease and Benign Paroxysmal Positional Vertigo, respectively. VN will be used for vestibular neuritis.

7   Practically all percentages like this will vary from article to article.

few hours, take a nap, feel tired for the rest of the day, then feel fine after that. Very few people with whom I spoke found this description to be accurate. One person who felt lousy for days after an attack was very relieved to read, in a book written by a nurse,[8] that many people experience lingering symptoms after an acute episode is over. This had been my experience as well. Many of those I interviewed also found the fatigue to last for days or weeks on end, often accompanied by minor vertigo or related symptoms. Furthermore, it does not seem universally true that attacks necessarily last only a few hours – sometimes they hang on for a day or more. Thus, quite commonly people present with symptoms that don't precisely fit the assumed pattern. It's not that anyone *wants* their symptoms to persist, but if what you read and are told is at variance with what is happening to you, you don't feel understood and may feel all the more in the grip of something menacingly mysterious.

In addition, there are a surprising variety of symptoms associated with vertigo, meaning different ways to experience the vertigo itself, to experience tinnitus, and to experience imbalance. It seems to be unclear what kind of importance to give these variations, if any, but there are many situations in which coming to a clear diagnosis is a challenge. I spoke with many who had never been given a firm diagnosis, or for whom it had taken a long while to arrive at one, or whose diagnosis had changed as things evolved over time. It is important to remember there is significant variation in the course of these disorders, and that for any given person vertigo may not recur or may respond well to treatment, or if it does recur may not pose a major problem.

Those with a diagnosis of Menière's or BPPV may suffer recurrent episodes of vertigo, which could be separated by weeks, months, or even years of feeling normal. Individuals with these diagnoses are not all affected to the same degree – for some it has major effects on their lives, for others it's more of a sideline. When I first was given the diagnosis, it would be an exaggeration to say that it felt like a death sentence, but to imagine being subjected to such attacks repeatedly certainly seemed a

---

8   Haybach, P.J.: <u>Meniere's Disease: What You Need to Know</u>

horrible prospect. I wondered how I could ever plan anything – a party, a vacation, a trip, a holiday – with any confidence. I'd never know when this was going to hit me again. I wondered about driving – who was to say I wouldn't be overpowered by a sudden rush of vertigo and lose control of my car? Even for those with relatively mild vertigo, the unpredictable timing of attacks, as well as the unpredictability of their effects, is a cloud that lurks in the back of one's mind.

Making the Diagnosis of Menière's Disease (or Syndrome)

The diagnosis of Menière's Disease (MD) is based on four "clinical" findings, meaning symptoms that can be observed by the clinician, or reported by the patient while being examined. These are sometimes presented in a tidy list, so that it looks as if making a diagnosis of MD is clear and precise, as follows:

- Attacks of vertigo, 20 minutes or longer
- "Aural fullness" or sensation of pressure in affected ear
- Tinnitus (usually confined to the affected ear)
- Fluctuating hearing loss in the affected ear

The first symptom, of course, is the presence of vertigo and its accompanying loss of equilibrium, or good balance. The vertigo may arise at any time, often suddenly and with no apparent precipitant. It may build up in its intensity, or arrive full force. Many people wake up in the morning with it. Most articles about vertigo state that it lasts for a few hours, followed by fatigue and the need for a nap. They go on to say that the person will feel well the following day. Less commonly, a source *will* note the vertigo may last a day or more. Length of vertigo attack (other than its lasting beyond 20 minutes,) type of vertigo, or experiences in the aftermath of an attack are not part of the diagnosis.

The second symptom – aural fullness - is a feeling of stuffiness or pressure in the head. The pressure sometimes extends to the face, and tends to be felt primarily on one side of your head. (This is because the disturbance in the inner ear is usually restricted to just one ear.) It resembles the pressure sometimes felt while flying. You may have the urge to keep swallowing or yawning in the hopes of making it disappear as you do on a plane, but it probably won't help. Some people say they feel as if their head is packed with cotton wool. Your own voice may sound distant or hollow to you. (Prior to the first onset of vertigo, I had had considerable pressure in my head and face all weekend, and assumed I was coming down with a sinus infection.)

The third symptom, tinnitus, is often called "ringing in the ears" and is something many people are troubled by even if they don't have vertigo. The ringing may be a minor background hum or it may be loud and disturbing. It may be a steady note, or it may rise and fall. It may sound more like buzzing or clanging than it does ringing. Some have described theirs as sounding like metal scrubbing against metal, some hear a flapping noise like a large bird in flight. Others say it sounds like whistling, or they hear more than one note at once. It may sound like the muttering of far off voices. For some people, it's so pronounced that it's almost impossible to ignore – it can interfere with sleep and with concentration, or even be so loud as to interfere with hearing. Like the vertigo itself, the presence and intensity of the tinnitus may vary. It tends to be worse during an attack, but may hang around at a lower level much of the rest of time. It's not apt to be extremely intrusive forever. In my case, it's a tiny high thin noise I'm normally barely aware of, so minor that I'm not sure when it first began. It may have started before I ever had vertigo.

The fourth symptom is a fluctuating hearing loss, meaning that it tends to improve between episodes of vertigo, and then get worse again if further attacks occur. For people with MD, the loss tends to be in the lower registers. This does not mean that patients necessarily become truly deaf – it means that their hearing acuity is reduced. But unfortunately, sometimes the loss is progressive, eventually leading to significant damage. Among the people I interviewed, there was considerable variation in the degree to which hearing loss was a significant issue.

Generally speaking, the two most significant features leading to a diagnosis of MD are the vertigo and the fluctuating hearing loss. Aural fullness, or pressure in the head, is not always present, nor is tinnitus. And many people who don't have vertigo are plagued by tinnitus. Of course many people are also hard of hearing – the difference with MD is that it fluctuates. But unless a patient is closely followed, the existence of this fluctuation may not be obvious. Since the loss may be mild, especially at first, you may not realize your hearing has been affected without a hearing test. For myself, I hadn't noticed much change in my hearing except in one specific way. When I held the phone to my left ear, people's voices sounded strange and it did seem harder to be sure of what they were saying. This has improved somewhat over time (maybe that's the fluctuating part) but has never returned to normal.

Because what's going wrong within the inner ear cannot be directly observed, a certain amount of conjecture is involved in trying to figure out exactly what's happening during a Menière's attack. It is believed to be caused by a buildup of excess fluid (**endolymph**) in the canals,[9] creating pressure that affects the tiny hairs (**cilia**) that communicate information about balance to the brain, via the **vestibulocochlear** (or auditory) nerve. In people unaffected by vertigo, the amount of endolymph is regulated by maintaining a balance between production and drainage. Somehow, in Menière's there may be either a blockage that inhibits adequate draining, or an overproduction of the fluid that creates the buildup. This condition is labeled "**endolymphatic hydrops**."[10] Because salt, in a general way, increases fluid retention, there's a prevailing belief that eating too much of it can precipitate an attack. Regardless of the role of salt, why the condition comes and goes remains mysterious.

Most doctors, especially ENTs, run a number of tests on people presenting with vertigo, particularly when MD is suspected. A hearing test probably heads the list. If no hearing deficit is found, you are less

---

9   More about this in Chapter 5, "Structure of the Ear."
10   Hydrops means excessive fluid within bodily tissues or cavities.

likely to be thought to have MD. If the hearing test showing loss is repeated at a later date and the results are changed, possibly even showing an improvement, that tends to confirm the diagnosis. You also may have an electronystagmogram (ENG) test, which measures vestibular function and tends to precipitate transient vertigo as well as nystagmus in people with MD. The direction, or beat, of the nystagmus will indicate which is the affected ear. A "caloric test," wherein warm or cold fluid is placed in the outer ear, may also be used to test the nystagmus. A videonystagmography, (VNG) which involves wearing special goggles, also tests the direction as well as the intensity of nystagmus.[11] These are all relatively simple tests that can often be done in the doctor's office. An MRI (magnetic resonance imaging) may be ordered to make sure there is nothing going on in the brain or nervous system that might be causing the symptoms.

Making the Diagnosis of Benign Positional Paroxysmal Vertigo

The cardinal symptom of Benign Positional Paroxysmal Vertigo, or BPPV, is a transient episode of vertigo elicited by some kind of movement, usually of the head. Some practitioners consider this the complete definition of BPPV, and in situations wherein brief bouts of vertigo is the primary complaint, the diagnosis seems clear. Tipping your head back, as you do when trying to get something off a high shelf or stare up at the night sky, is a motion that commonly sets off an attack. Lying down or turning over in bed is also a frequent trigger.[12] The reason the condition is called Benign *Positional* Vertigo has to do with the fact that the motion and position of your head contributes to or provokes the vertigo. The word "paroxysmal" may refer to its brevity. I presume it is called "benign" because it's not fatal. The relationship between specific motions and a burst of vertigo is particularly evident when attacks are brief.

---

11 This is not a complete list of all the possible tests that may be administered.

12 People suffering with MD or VN may also have some of these "positional" reactions, but they are not usually the primary precipitant for full-blown vertigo attacks. These crossover symptoms may sometimes make it hard to settle on a diagnosis.

However, in many cases wherein people have received the diagnosis of BPPV, intense attacks of vertigo (sometimes spinning, sometimes the world moving) are experienced in just the same way as they are in Menière's and may last for hours, accompanied by the same appalling trouble with balance. Among those I interviewed,[13] most of the people diagnosed with BPPV were subject to the longer lasting episodes. Many experienced absolutely overwhelming symptoms – persistent vertigo, on the floor unable to stand, vomiting. Others had vaguer feelings of being sick, dizzy, off-balance, and might be able to walk unaided, albeit with difficulty. In terms of the diagnosis, what is specifically different about BPPV is that tinnitus, pressure in the head or ear, and hearing loss are not included. Something that can muddy the waters quite a bit is the fact that many people with the diagnosis of BPPV do have tinnitus, hearing loss, or both, just as do people in the general population – but these are not attributed to the vestibular disorder.

People who experience the short bursts of vertigo often get them in clusters, meaning there will be a few days when they are subject to them, then long periods of time will go by without them. Anyone who has had a lengthy attack may envy the person for whom it is only minutes, but vertigo is always distressing, and several short attacks during the course of a day is a disconcerting experience.

Ursula's experience may help to illuminate this. Beginning in her mid-fifties, she would find that lying down in bed would precipitate a sudden sensation of spinning, lasting less than a minute. During the day, moving her head in certain way would do the same. For a week or two, this would happen many times a day. Then it would all be over, and might not come back again for months. Although when it first occurred, she was frightened and wondered about a possibly ominous diagnosis, she was reassured by a visit to her internist, who told Ursula she thought she had BPPV. Partly because of the brevity of the attacks, and the infrequency of the times she was susceptible to them, they seemed mainly a nuisance once she knew they weren't life threatening.

---

13   Remember that my sample is random. I interviewed anyone who responded to my request, so that the sample is not representative of trends in the general public.

Ursula also figured out the positions and motions that triggered the vertigo, and thus was able to avoid much of it and feel in control of the condition. Katya and Eva, who also have short-lived spells of vertigo, do not have the clusters of attacks that Ursula has, but from time to time will suddenly experience a short burst of it in response to a particular head movement. All of these people have short episodes, and experience their condition differently than do people whose vertigo lasts a long time.

Current understanding of the cause of the condition is quite different from that of MD. It is believed that tiny little "rocks" or calcium crystals[14] (**canaliths,** or **otoconia**) move out of place within the inner ear, and disturb the cilia so that faulty information is communicated to the neurons leading to the brain. When in their proper place, these otoconia[15] reside in the **otolith organs,** called the **utricle** and the **saccule.** The otolith organs are situated at the base of the three semicircular canals, where they all come together. Usually when they break free they float in the semicircular canals, typically the posterior. But sometimes instead of floating around they get stuck in the sense organ (the **cupula**) at the base of one of the canals, which also causes vertigo. This condition is called **cupulothiasis.**[16]

When a patient experiencing vertigo goes to see a doctor, s/he will be likely to want to check first for BPPV, since it is much more common than MD.[17] A simple test called the Dix-Hallpike Maneuver is one way to do this. The medical practitioner will put the patient through a series of motions, involving turning their head, lying back, sitting up. He or she will observe the eyes to check for the presence of nystagmus. If it is present, the type of nystagmus will indicate which ear is the affected one. It will also tend to confirm the diagnosis of BPPV and rule out the diagnosis of MD, since the maneuver doesn't have much effect

---

14 Sometimes jokingly referred to as "rocks in the head." However, these "rocks" are likely to be miniscule, smaller than most grains of sand.

15 When in place, the crystals are called otoconia, when they're out where they don't belong, they are called canaliths.

16 More information on the structure of the ear will be found in Chapter 5.

17 About 2.4% of the population will experience BPPV at some time in their lives, according to the Vestibular Disorders Association (VEDA)

on people who don't have BPPV. Unfortunately, as of this writing many emergency rooms and private practitioners are unfamiliar with the use of this maneuver, which can lead the patient on a protracted search for help. The motions employed in the Dix-Hallpike are similar to those of a somewhat more elaborate procedure, known as the Epley Maneuver,[18] that is a highly effective treatment for BPPV, about which more will be said later. I should note here that, if the medical facility where you go is familiar with the Epley Maneuver, they may skip the Dix-Hallpike and go ahead with the Epley right away.

Making the Diagnosis of Vestibular Neuritis

Vestibular Neuritis (VN), also known as neuronitis, is thought to be a viral infection of the vestibular nerve and/or the vestibular nerve ganglion,[19] found just outside the vestibule. The type of virus could be either herpes or varicella zoster. It presents with an acute spontaneous vertigo, similar to what occurs in MD, but unlike MD it does not damage one's hearing. Also unlike MD tinnitus is not part of the diagnosis. As with the other vestibular illnesses, the severity of the vertigo and imbalance will vary from person to person. According to the Tampa Bay Hearing & Balance Center website,[20] the vertigo is often described as a perception of movement oscillating side-to-side or tumbling. Constant at first, the patient may need to lie still in a dark room. As with MD, there may be nystagmus, nausea, vomiting, exhaustion. The "head thrust test" may help to distinguish VN from the other vestibular illnesses. Normally, in turning one's head to the side, your eyes tend to stay centered. But if they move to the side along with your head, as a doll's would, this is indicative of VN. However, like a lot of these tests, the "doll's eyes" phenomenon is not always present nor is it always restricted to VN.

According to the website of the Chicago Dizziness and Hearing,[21] the patient may experience the acute stage of vertigo nearly constantly for

---

18   There are other similar maneuvers that particular doctors or other practitioners may prefer, such as the Semont or the Gans Maneuvers.

19   This ganglion consists of a group of the cell bodies of the vestibular nerve.

20   Tampa Bay Hearing & Balance Center, "Vestibular Neuritis & Labyrinthitis."

21   dizziness-and-balance.com

many hours, or even days or weeks, before it gradually recedes. Thus, the initial attack of vertigo due to VN can in some cases be the most prolonged of these four disorders. During the recovery period, as the symptoms subside, the individual is likely to have brief moments of vertigo, often induced by movement, similar to that found in BPPV. During that time, people may also have trouble concentrating, feel fatigued, foggy-headed, or disoriented. VN sometimes returns later on to haunt you, but is less likely to recur than MD. According to a large study conducted in Taiwan, 40% have at least one recurrence, and of that group 38% have two, and a small number have three or more. That would indicate that only about three percent of all those afflicted with VN are beleaguered by numerous repeated episodes.[22] However, an unfortunate possibility is that the nerve may have sustained some damage, leaving the patient with lingering symptoms of disequilibrium or of more minor dizziness.

The main thing that distinguishes VN from MD at the beginning is the fact that one's hearing is not affected. And as the person returns to health and gets on with his or her life, the greatest likelihood, unlike MD, is that the unwelcome experience of vertigo is over, or if it does recur, it's not likely to do so again. VN and MD are both less common than BPPV.

Making the Diagnosis of Labyrinthitis

Like MD and VN, labyrinthitis tends to start with a sudden acute episode of vertigo. Like VN it is an infection, also usually caused by a virus. Less often, an attack of labyrinthitis may be secondary to a bacterial middle ear infection, in which case it can be treated with antibiotics. Unlike VN, but like MD, one's hearing may be affected, but it does not fluctuate as it does with MD. The intense vertigo may last for hours and sometimes days, and then fade away. Full recovery may be slow, sometimes lasting weeks or even months.[23] As it recedes, there may be continuing but less severe symptoms. The main concern is the possibility of ongoing hearing impairment. But except for damage that

---

22 Tampa Bay Hearing & Balance Center, "Vestibular Neuritis & Labyrinthitis."
23 Johns Hopkins Medicine Website. "Labyrinthitis." See bibliography for URL.

may have occurred, when the patient has recovered that will be the last of it, as with most infectious diseases. It won't continue to be an ongoing scourge in one's life. If the person does suffer a recurrence of vertigo, the chances are that he or she has MD, after all.[24] Diagnoses aside, several people have told me of themselves, or a friend or relative, who had suffered a severe bout of vertigo for a few hours or a day that disappeared afterward. Perhaps these were due to labyrinthitis or vestibular neuritis, perhaps due to some mysterious and undiagnosed condition.

End Note

Articles about the diagnoses of VN and labyrinthitis state specifically that a patient can expect a recovery period as symptoms fade, wherein they experience minor symptoms of vertigo, along with some fogginess of mind or disorientation. Less tends to be said about this with regard to MD or BPPV. Yet, as noted above, people suffering from either MD or BPPV may also experience days or weeks of symptoms that interfere with functioning following an attack. One of these symptoms is of ongoing fatigue. Another is impaired concentration, which may be especially noticeable when trying to "multitask," or do a number of things at the same time. A challenge any time in our busy world, this can be a particular problem at work. The individual may feel woozy, or a little "out of it" for much of the day. Various minor vertigo-related symptoms may be a source of distress from time to time as well, such as brief bursts of wooziness or faint feelings, zingy sensations in one's head, noises in one's head, and so forth. Balance, too, may take a while to resolve – people may have trouble walking a straight line even if their heads feel clear. None of these is included as such in a diagnostic picture, but they are part of the experience for many who have vertigo, no matter the diagnosis.

---

24   www.dizziness-and-balance.com "Labyrinthitis and Vestibular Neuritis"

4

Sources of Diagnostic Confusion

Although some people suffering from vertigo and its related symptoms are given a diagnosis reasonably quickly, many are not. Many people remain overwhelmed and confused by their symptoms for years. Sometimes they may have a diagnosis, but it makes little sense to them. Sometimes the diagnosis itself is a source of despair, especially when the treatments offered don't seem to make much difference. Many with vertigo are given one diagnosis at first, another later, sometimes a third one after that. In the previous chapter, I mentioned that a variety of diseases and conditions can precipitate episodes of vertigo, so figuring out what to rule out can pose a diagnostic challenge.

Migraine Associated Vertigo

One of the most significant crossover possibilities is migraine-associated vertigo, abbreviated to MAV. The incidence of migraine is more common in people with MD (Menière's) than it is in the general population. Overall, studies show that while 13% of the population has migraine, among those diagnosed with MD, 50% do, according to the website of Timothy Hain, MD.[25] He sees migraine as being the most common source of vertigo, somewhat surprising to me since no

---

25   www.dizziness-and-balance.com

one I interviewed brought it up, and no doctor who examined me inquired about it – at least not that I remember. Migraine, like MD and BPPV, is a remitting condition. Hearing loss is not likely to be part of MAV, or if hearing is affected, it will usually be in both ears. Nystagmus doesn't tend to be part of it either, or if present it's more likely to be vertical (going up and down) instead of horizontal (zipping back and forth.)

Oddly, a person with MAV doesn't always have a headache or pain – the vertigo may be a substitute of sorts. Clearly, this is another one of those things that can make arriving at a diagnosis confusing. Having MAV means that your medical provider may be able to treat you with drugs used for the treatment of "regular" migraine. As with MD, some dietary restriction would be tried first – avoiding foods that generally tend to trigger migraines, such as MSG, chocolate, cheese, alcohol. There are various choices as to the drugs themselves. Common ones are the tricyclic antidepressants, amitriptyline in particular. These are the older types of antidepressants, in use before the development of Prozac and its ilk.[26] Their effect on migraine is idiosyncratic, meaning the medications were developed for treating mood disorder, but here they are treating headache. A class of drugs initially developed to treat various cardiovascular symptoms may also be prescribed. These are beta blockers, calcium channel blockers, and medication for hypertension. Or antiseizure medication, such as Topomax (topiramate) may be used. Recognizing that migraine itself is a remitting condition, nonetheless these – and other – treatments for migraine open up new avenues for help for the person whose vertigo is part of a migraine syndrome.

Acoustic Neuroma

An **acoustic neuroma** - a small non-malignant tumor on the vestibulocochlear nerve also presents with vertigo but no hearing loss. It can sometimes be detected on an MRI. It is sometimes treated surgically, sometimes with radiation, sometimes left alone. It is rare. There

---

26   These would be the selective serotonin reuptake inhibitors, or SSRIs.

are a host of other conditions that cause a similar set of symptoms, but have their origins outside the inner ear.

Atypical Menière's Disease

MD itself may have "atypical" forms.[27] One is called "cochlear MD." This presents with tinnitus, episodic fullness or pressure in the head, as well as a fluctuating hearing loss – but no vertigo!! Sometimes the vertigo comes later, possibly after a few years. Once it emerges it would no longer be called "cochlear" - it will have morphed into plain old Menière's. The other atypical is "vestibular MD," which has vertigo, tinnitus, head pressure, all fluctuating - but no hearing loss. At the current time, making these diagnostic distinctions probably doesn't change treatment recommendations – they would be the same as for "regular" MD, especially at first.

An example of an atypical presentation that was eventually diagnosed as Menière's is Ethan's saga, which did not begin with vertigo. Instead, he woke up one day with a very loud "rumbling, a high-pitched roar" in one ear. It lasted weeks, with varying intensity, and was extremely uncomfortable – "an incredible distraction." He did see a doctor within the week, and had an MRI and a hearing test. The MRI found nothing amiss in his brain, but the hearing test did show loss – although Ethan thought it was the noise of the tinnitus that made it hard to hear. No diagnosis was made at the time. Steroids and antibiotics were tried, to no avail, but after six or eight weeks, the tinnitus began fading. He continued to have episodes of it, accompanied by nausea and funny feelings in his head, at least twice a week. He tried acupuncture, which he thought helped a little. Overall, he didn't feel the situation was understood, it seemed his doctor was at a loss, it seemed there was no end to it. After 18 months, he contacted a new doctor who diagnosed him as having MD, which gave him a "huge feeling of relief." Ethan found the dietary restrictions that are typically recommended for MD to be very helpful. Seven years after the onset

---

27   Menière's Disease Information Center (www.menieresinfo.com)

of symptoms, Ethan had an attack of disabling vertigo that lasted for a day and a half. It was severe enough to keep him in bed, especially since he was unable to move without vomiting. Over all those years and since, this is the only attack of actual vertigo he experienced. He says he has not had head or ear pressure, and no hearing loss. The difficulty in reaching a diagnosis of MD here seems clear: Ethan was not initially complaining of vertigo, nor really of anything on the list of classic MD symptoms, except for tinnitus.

More Diagnostic Confusions

Blows to the head, even those incurred years ago, can lie behind episodic vertigo, sometimes in the form of BPPV, and are not always recognized as a cause. Occasionally BPPV can be precipitated by dental work or surgery, wherein the person's head is kept back in the same position for a prolonged time. This happened to at least two of my subjects, and for one of them the dentist vehemently denied the visit could have had anything to do with the vertigo. Theoretically, in these situations the vertigo would not be accompanied by tinnitus, hearing loss, or aural fullness.

To recapitulate, the diagnosis of MD includes a set of four symptoms, but it is possible for a person to start off with tinnitus or hearing loss and develop vertigo later on. For BPPV, the symptom of vertigo alone leads to the diagnosis, yet these people may coincidentally suffer from tinnitus and hearing loss. The fact that labyrinthitis and MD both present with vertigo and hearing loss also presents a challenge to the diagnostician.

Among the people with whom I spoke, about a third were given a diagnosis within a short time after their first contact with a doctor. Another third were never given a diagnosis, or had an array of symptoms that didn't fit neatly into any one diagnosis, or had gone through a progression of diagnoses, for example BPPV at first and later on MD. Sometimes this seemed to be attributable to the complexity of their symptoms – they didn't fit well into a specific diagnosis.

Certainly many people with MD will occasionally experience vertigo as a response to changes in position, even though sensitivity to position doesn't play as major a role in their condition. The classic one is lying down or turning over in bed. A small group of people with whom I spoke had relatively mild symptoms that sounded to me like BPPV, and were given some directions for coping with it, but not a diagnosis. Possibly because these folks were not too troubled by the condition, they seemed comfortable with thinking of it as just "vertigo."

In addition, there was another group among my interviewees who were not given a diagnosis for a long time – months or years – and for whom this fact added significantly to their distress and anxiety. They felt themselves adrift on stormy seas with no sense of direction, no guiding compass. Evidently they appeared to present diagnostic mysteries to the medical personnel they consulted as well. Some such stories are presented below, some will appear elsewhere in the text. For a patient, the lack of clarity as to diagnosis can play out into long and bewildering stretches of time, time when life may be deeply affected by bouts of vertigo, time wherein the individual feels at a loss. Or the patient may have a diagnosis, but find his own experience doesn't match what he's been told, again leading to the feeling of floundering.

Jody's saga actually began before she developed vertigo, since she had had tinnitus all her life. In her early 30s, at a time when she was living with a roommate, she suffered a vertigo attack that lasted for days, such that she was forced to stay in bed all day, leaving it only to crawl back and forth to the bathroom. Sometimes she lay on the bathroom floor for a long time, vomiting often, sucking on ice cubes to stay hydrated. She described the vertigo as her "eyes bouncing around" making it hard to see. "If I had to look up, I was like a mole coming out from underground." Tinnitus was loud and her ears hurt. When she was able to stand, her fiancé took her to see a doctor, who thought she had an ear infection and prescribed antibiotics. Her symptoms improved somewhat, but the tinnitus grew louder and the pain in her ears continued. She experienced several "mini-whirls" a day, forcing her to sit down until they passed. And she was exhausted all day long – it never left her. This went on for a few weeks, so she returned to the doctor who again prescribed antibiotics, although of a different sort and in a higher dosage. The medication didn't make any noticeable difference, but she decided to return to work and see if she could manage. She tried to disguise the times when the vertigo hit her by looking down and sitting very still. She remained fatigued, and found her concentration was affected. Life went on this way for a while, feeling poorly much of the time but hanging on, until another major attack hit. At that point, Jody contacted her doctor again and insisted on being referred to a specialist at a nearby university hospital. There she received a full work-up, and was given the diagnosis of MD. What can be seen here is that her internist was not thinking in terms of a disorder of the vestibular system, or inner ear. Possibly because her ears ached, he seemed to assume that it was an infection. Even though little improvement was seen after the first course of medication his assessment didn't change, as evidenced by his prescribing antibiotics all over again.

Caleb began experiencing symptoms, also diagnosed at the time as ear infections, in his early 30s. This was about 20 years ago, around 1994. He had pressure in his head, tinnitus, and noticed a mild hearing loss – but no vertigo. That did not begin until another two years, and when it did, it was mild at first. Despite its being mild, it was more pronounced when he was in or around moving things. At the time he drove a truck for a living, and thus was constantly dealing with motion. He ended up getting hurt falling off the back of his truck due to his poor balance. He was frightened by his symptoms and worried they might be due to a stroke. From the beginning, he'd been in touch with his internist but no conclusion had been reached about what was wrong. In 1999, five years after his initial symptoms and three after the onset of vertigo, he was referred to an ENT who administered a series of tests, as well as a CT Scan. A lab technician called out, "You have MD," and he felt "scared to death," fearful this meant be would become deaf. This story tends to sound as if the diagnosis could have been straightforward and easy to reach – he had all four basic symptoms of MD - but somehow it took several years to get there. As with

Jody, the fact he was initially thought to have an ear infection delayed diagnosing what was actually wrong. (It's also possible that the head pressure and deafness were mistaken for infection.) But once the vertigo itself began, still three years before the final diagnosis, it seems surprising that didn't clarify things.

Thus, in the group of people I interviewed, it was not uncommon for people to wait a long time before getting a diagnosis. A few with very brief vertigo were neither diagnosed nor treated. Other than those, most – but not all – of the people with BPPV received their diagnosis soon. This was not true of those with a diagnosis of MD – it was the opposite. Many did *not* receive this diagnosis after first contacting a doctor, and in several cases it was a few years before the diagnosis was settled on. In this regard, I was lucky, not that I embraced the notion of having MD cheerfully.

5

Structure and Functioning of the Ear

Functional Basis for Vertigo: How it All Seems to Work

Learning more about vertigo leads to an appreciation of the workings of the vestibular system within the inner ear, and the recognition of the astounding complexity that lies behind all of our senses. Despite the fact that knowledge about our bodies and disease has mushroomed over the past 100 years, so much is still remote and mysterious to us. Gaining solid information about the inner ear is impeded by the fact that it is buried deep within the temporal bone near the brain, inaccessible to peeking or prodding from the outside. Its workings are both intricate and miniscule, with each part depending on the smooth functioning of other tiny little parts and systems. When the vertigo and its accompanying imbalance are caused by a disturbance in the functioning of the inner ear, it is referred to as "peripheral" vertigo, to distinguish it from conditions arising in the brain or nervous system. The latter are labeled conditions of "central" origin.

Your ear is divided into three sections: the **outer ear**, which is the external part where the sound is gathered, consisting of the ears that we see on the sides of our heads (called **pinnae**), as well as the ear canal that leads inside. The ear canal terminates with the eardrum (**tympanic membrane**), which creates a complete seal across the canal. Sound, which is carried through the air on waves, pulsates against the eardrum, causing it to vibrate. The fact that the eardrum is situated well inside the head is clearly a protection, and may help to focus the sound.

Next is the **middle ear** (called the **tympanum**[28]) where the sound waves are transformed into impulses to be transmitted to the brain. It is located within the temporal bone. The pulsating movement of the eardrum activates three tiny bones (**ossicles**), known colloquially as the **hammer**, the **anvil**, and the **stirrup**, and scientifically as the **malleus**, **incus**, and **stapes**. The hammer is attached to the eardrum, and communicates with the anvil, which in turn communicates with the stirrup. (Think of a blacksmith whacking a hammer on an anvil in order to shape metal into a stirrup.) The stirrup (stapes) connects with the "round window" – another membrane – through which the impulses are sent to the **cochlea**, the part of the complicated inner ear that gives us hearing. However, the **middle ear** itself is a relatively empty space, and is connected to the back of the nose and throat through the **Eustachian Tubes**. These keep the pressure exerted by the atmosphere equalized on both sides of the eardrum. Most people notice an increase in such pressure when you are up in a plane, or diving down deep in the ocean. When you have an ear infection, the middle ear is usually where it is located.

The **inner ear**, by far and away the most complicated, has two sections, all nestled within an organ the size of a grape!! (*In attempting to describe it, I am oversimplifying in the hope of offering a schema that can be readily understood*). The **cochlea** receives and sends information about *sound* to the brain, while the **vestibular system** regulates *balance*. The vestibular system lies within a space in the bone called the **vestibule** and includes the three **semicircular canals**, along with organs called the **saccule** and **utricle**. Despite the difference in function – one hearing, one balance - these systems are intimately connected.

---

28   Notice that the root of the word is the same as for the tympani – the large drums – of an orchestra.

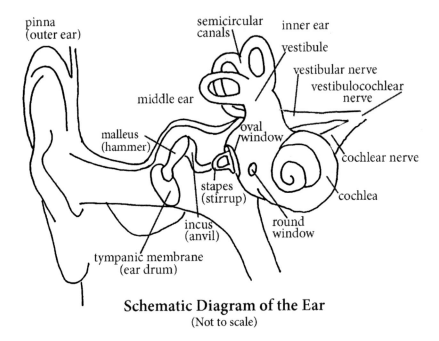

Schematic Diagram of the Ear
(Not to scale)

The cochlea resembles a spiral snail or conch shell, perhaps the size of a pea, and is a coiled bony canal or tube, which contains a fluid called **perilymph**. The canal is lined with miniscule hairs called **cilia**. After the sound impulses are transmitted through the hammer and anvil, the stirrup delivers them to the **oval window**, which is at one end of that spiral tube. This motion activates the perilymph, which in turn moves against the cilia, like currents causing underwater plants to sway back and forth. The cilia convert the motion to electrical signals that communicate via **neurotransmitters** to nerve receptors, then on to the **vestibulocochlear**[29] **nerve**, and thence to the brain. It's a kind of chain reaction in a very tiny and complex space. A membrane called the **round window** is situated at the end of the canal, and bulges in and out in response to pressure from the perilymph.

The vestibular system, the seat of balance, is comprised of the three

---

29   Sometimes referred to as the auditory or acoustic nerve, especially in older publications. The nerve has two parts – the cochlear for the auditory signals, and the vestibular for the balance signals.

interconnected semicircular canals,[30] each consisting of an inner membranous tube surrounded by a bony outer one. The canals look like loops, and are positioned roughly at right angles to each other. They also are lined with cilia (minuscule hairs), and filled with a fluid known as **endolymph**, different in chemical composition from the perilymph. When your body moves, the endolymph moves as well, stimulating the cilia, which in turn transmit information to your brain via the vestibular, or **vestibulocochlear**, nerve, enabling you to get around smoothly. The movement of the endolymph is dependent on gravity, so that when you lean down, the lymph slides, just as you would expect it to do. This can cause trouble for astronauts heading for the stars in the gravity-free environment of their spacecraft – there is no up, down, or sideways without gravity's pull. With the three canals placed at different angles, the endolymph will be moving in slightly different directions, so as to send a full report of your body's position to the brain. When you think about it, you can see that the smallest variation in motion is detectable – whether you are leaning over to tie your shoes, hanging upside down on a trapeze, or staring up with your head cocked toward the ceiling. The canals are sensitive to rotational movements – meaning the turning of your head - in order to maintain equilibrium while you run, whip around corners, or simply stroll along a sidewalk, weaving your way around other pedestrians. If the system is working well you will maintain your balance and your feeling of being centered, no matter what you are doing, and never give it a second thought. And it works instantaneously.

Between the semicircular canals and the cochlea, and connected to them, is a small area which contains the **otolith organs** called the **saccule** and the **utricle** (mentioned earlier in connection with BPPV.) These are positioned at right angles to each other, so as to detect horizontal and vertical motion, as well as your relationship to gravity. They are sensitive to the acceleration of linear motion, meaning the speed at which you are moving straight ahead, as well as its opposite:

---

30   The three canals each have a name: the anterior (at the front), the posterior (at the back), and the horizontal.

deceleration. You will be jogging along and come to a sudden stop and they are right there, monitoring your balance. The otolith organs contain a gelatinous substance in which tiny crystals of calcium carbonate (the otoconia)[31] are nestled. As with the semicircular canals, very slight variations in motion and position will stimulate them, which in turn stimulate cilia that are located in the cupula where the sense organ lies. The cilia convert the movement to electrical signals to be conveyed to the vestibular nerve. Very soon after the vestibular nerve leaves the cupula it joins its fellow, the auditory nerve, to form the vestibulocochlear nerve.

The vestibulocochlear nerve sends the signals it has received, with regard to both hearing and balance, to the back of your head where the **brainstem** is situated.[32] This is the part of your brain where all the nerves serving the head (the **cranial nerves**) connect. It adjoins the top of the spinal cord, and all the nerves - those going from the body to the brain (sensory nerves) and those going from the brain to the body (efferent nerves) pass through it. The brainstem has little to do with the thinking, conscious part of your brain. Instead, it controls most of the workings of your body that are outside your awareness, including cardiac, respiratory, and digestive functioning, all the things that chug along day and night, keeping you alive.

Think about a common example, that of a child twirling around with the intention of producing vertigo. What probably happens is that the endolymph gets started spinning too, and doesn't stop right away when the child flops to the grounds. Thus, for a few seconds, the child will continue to feel as if he is spinning until the endolymph quiets down. In the case of vertigo caused by a malfunction of the inner ear, similar signals are being sent to the brain so that you respond as if the endolymph were swirling or moving. Something is happening that causes the cilia to give faulty information so that the sensation of spinning may persist for agonizingly long periods of time.

---

31   As noted previously, it is these tiny crystals, known as canaliths when they move out of place, that cause the symptoms in BPPV.

32   It is worth noting here that the vestibulocochlear nerve is a sensory nerve, meaning that it carries information from the senses to the brain. It is the 8th cranial nerve.

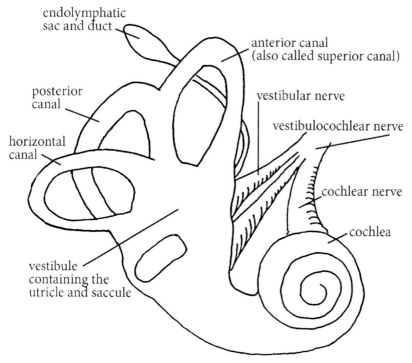

Schematic Diagram of the Inner Ear
(Not to scale)

One of the amazing things about balance and vertigo is the close connection this system has to your eyes. Anyone who has suffered an attack of vertigo is aware of how sickening it is to look around as the area surrounding us appears to be moving, twirling, swinging. And we know that our eyes are, in fact, jerking back and forth with **nystagmus**, or **saccadic movements**. For a long time, I wondered about this. I couldn't see exactly what balance had to do with jerky eyes, and wondered if there was a vestigial evolutionary relationship, maybe something left over from our long long ago fish ancestors, something they needed as they prowled through the primordial soup. To my great interest, it turns out there is a connection called the **vestibulo-ocular reflex** (the VOR.) After the information from the inner ear arrives in the brainstem via the vestibulocochlear nerve, three cranial nerves[33] carry it out to your eyes. Normally the function of this connection is to keep your eyes centered, no matter which way you turn your head. If you are looking at an object or looking off in the distance, and you move your head to the left, say, your eyes will automatically move to the right to keep your gaze just where it was. In this way your view of the world is stabilized through the interaction of eyes and inner ear. It would appear that when your vestibular system runs amok and tells your brain you are whirling about (when in fact you are sitting stone still) your eyes are frantically trying to catch up, whipping back and forth as the inner ear emits a signal that you are in rapid motion.

Thus, there is clearly a connection between your eyes and your vestibular system. They work in concert to help keep your view of the world steady, but the impulses and information run back and forth through the brainstem. You need this functional connection to be oriented to the world. But despite the fact they are nestled together in the same tiny organ, there does not appear to be much functional connection between the vestibular system and the auditory one. When they do have an effect on each other, it does not seem to be a very good one. Remember that repeated attacks of vertigo in MD are often repeated attacks on your hearing as well. And people prone to vertigo are often hypersensitive to sound.[34] Even if some of their hearing is diminished, they still may find it hard to tolerate loud noises, to the extent that they experience actual pain. In some cases, intense sound can actually precipitate an attack of vertigo.

There is also an interconnection between the vestibular system and the body that extends into the neck and on to the rest of the body, with those in the feet being particularly important. Nerve fibers feed information about motion and position to the central nervous system. The inner ear responds and automatically works to maintain posture and stability even as people jump, climb, stumble over uneven terrain, or are thrown off balance. Sensors that operate *within* the body, and thus send internal

---

33   They are #3 (ocolomotor) #4 (trochlear,) and #6 (abducens.)
34   This sensitivity is called hyperacusis.

information from one part to another, are called **proprioceptors**. (This is distinct from the familiar sensors that pick up the *external* stimuli in our environment, such as light, sound, odor, etc. – our "five senses.") Note that the response from the vestibular system to this internal stimuli is mechanically different from those reviewed above, wherein the vestibular system reacts to changes in fluid movement in the inner ear itself. In this case, information arrives in your brainstem via nerve impulses fired up *outside* the vestibular system. Without proprioception in your feet, you would not be able to walk without paying conscious attention to the placement of your feet, and would be challenged by slanted surfaces, brick sidewalks, anything that wasn't perfectly flat and smooth. Proprioception can be disturbed by inner ear disorders, as many people find themselves walking on a floor that feels to them like a rubbery unstable surface, or are thrown off balance by the slight slant of a ramp.

In the end, it is the brain, the recipient of those messages that sometimes make us feel so sick, that interprets the messages and directs balance.

Other Possibilities Concerning What's Happening in MD

In Chapter Three, I presented the malfunctioning of the inner ear in the way you are most likely to see it described. For many years – better than 70 – people have believed that endolymphatic hydrops (ELH) is the root cause of vertigo and the other symptoms associated with MD. On autopsy, the temporal bone, where the inner ear is situated, shows changes indicative of hydrops. In addition, experiments with animals show that when ELH is artificially created, vertigo develops. People who have MD have been shown to have ELH on post mortem examination, but it turns out there are some who have ELH but no history of MD.[35] An MRI using a special contrast material has made it possible to observe ELH in living subjects, again confirming the presence of hydrops not only in MD, but in unaffected persons as well.[36]

---

35 Berlinger, Norman; Menière's Disease: New Concepts, New Treatment, Minnesota Medical Association, November 2011

36 Pykko, Ilmari et al; Menière's Disease: A Reappraisal. BMJ Open e2013 001555 vol.3 issue 2

Over the years, there has been ongoing debate concerning whether a dysregulation of the composition of fluid within the inner ear is the culprit, either on its own or in addition to ELH.[37] There are those who believe that excessive pressure within the canals causes tiny ruptures (referred to as "pinhole leaks") in the membranes that line them, so that the endolymph mixes with the perilymph, creating a chemical change that is responsible for the faulty message relayed to the brainstem. A chemical change would be a different kind of explanation from the explanation that physical pressure against the cilia, created by an overabundance of endolymph, is the source of vertigo in MD. A couple of the people I interviewed thought of their vertigo as caused by pinhole leaks.

It is interesting to see there are some who believe that MD and BPPV are part of the same condition.[38] I mention this ever so briefly since it helps to confirm that investigation is ongoing into vestibular illnesses, that researchers reflect upon accepted ideas, and that there is recognition that there is much to be learned.

An important reason for studying the structure and physiology of the inner ear is to gain understanding that could lead to effective treatment. Over the years, a large volume of information has built up, a great deal of it available on the Internet. Much of what is to be found there is repetitive and some of it is conflicting. Scientific studies, even those with a respectable sample size, don't always support the same conclusions. So far, none of the findings and ideas challenging the primacy of ELH as the cause of MD has led to new kinds of treatment. The intermittent nature of vertigo attacks remains a mystery. Given the common agreement, which was originally based on autopsy studies, that the inner ear is overfilled with endolymph, does it drain out and go back to normal between acute episodes, or is it that you somehow become tolerant of it? After having been quiet for a while, what makes the symptoms come back?

---

37 Rauch, Steven; See Bibliography.
38 Phillips, J & Prinsley, P; "A Unified Hypothesis for Vestibular Dysfunction?" <u>American Academy of Otolaryngology</u>,(2009) 140, 477-479

6

Vertigo Drags On – My Illness

My Illness Over the First Few Months: February-December 2006

After the first weeks it seemed pretty clear, at least to the doctors, that I had Menière's Disease. Although I could see how this conclusion was reached – I met all the criteria for the condition (vertigo, hearing loss, aural fullness, tinnitus) – I resisted accepting the idea for a long time. I didn't like the implications of it – being someone who might at any moment collapse out of control, or a frail and pathetic person grabbing hold of furniture to steady myself as I felt my way around a room. I kept thinking of a story I'd read ages ago, where a kid would say in a dismissive or disgusted tone, "Ma's having one of her spells again!" and wondering if that character had vertigo, and not wanting to be that Ma. I clung to the idea that, even if I *did* have MD, it would go away and not return for years, preferably never. The first spring held little promise of that, since I had precious few days where I felt entirely normal.

During each month during that spring, there would be at least eight days marred by an episode of vertigo, but never as bad as the first time. Sometimes the dizziness would seem to pour into my head and I would be unable to look directly at people, and need support in order to stand or walk. But I wasn't on the floor and I wasn't throwing up, despite feeling ill and nauseated. I'd feel lousy for a few hours and then it would float away, leaving me exhausted. On the days when I didn't have actual vertigo, I often felt significantly fatigued, and usually experienced sensations associated with it, like pressure in my head, or tinnitus, or clicking in my ears, or brief bursts of dizziness that struck without lingering, or a more prolonged woozy out-of-it feeling. A great deal of the time my balance was poor, such that it was hard to round a corner – I would just keep on going along my original path, or stumble as I attempted to change direction. Because of the frequent nausea, I lost about ten pounds. Most of the time, I went to work, but it was a struggle. I probably felt well around half the time, but despite that it seemed as if this new and distressing sensation dominated my life.

During those months, I occasionally took Meclizine, medication suggested by the doctor to tone down dizziness. It is a drug similar to Dramamine, typically used to control seasickness. I never found that it made much difference, but there are many people prone to vertigo who swear by it. I also took Valium from time to time, medication normally prescribed for anxiety that is thought to quiet the inner ear. It seemed to do a bit more good than Meclizine, but neither one was remarkably helpful. Both drugs had the downside of making me feel drowsy and tired, and I was tired enough already. Again, there are many who have a better experience with Valium. Except for needing to sleep a lot, there didn't seem to be much I could do that made any difference. However, things became definitely better by around mid-July – I was able to do some hiking on a trip to Arizona, and enjoy swimming and surfing during August. I did have brief daily reminders of my condition in various forms: pressure in the head, passing wooziness, feeling like my brain was sliding around, clicking noises, and so forth – but not all at once, and not all day long, and best of all, not accompanied by fatigue. By late summer, my energy picked up and I was pretty normal most of the time, although still with some form of daily reminder that all was not quite as it should be. During all the last half of that year, the fact that something was wrong with me was not likely to be noticed by others.

During those first few months, I had a few doctor's appointments early on. Apart from the one I'd seen in the emergency room, I met with the ENT who gave me a thorough evaluation, including a hearing test and other standard tests. Eventually, four months after the onset of vertigo (in June) he ordered an MRI. The MRI confirmed there were no anomalies in my brain, and no tumors that might be causing the trouble. In the way of treatment, I was given a course of steroids orally during the very first week, and later on after a flare-up. My symptoms did ameliorate during the week I was taking them, but it was not clear to me that the pills were making the difference, since the symptoms rose and fell so much anyway. I was prescribed a diuretic in the hope it would help to reduce pressure in the inner ear. In addition to the Meclizine and Valium, I also tried a B vitamin that was supposed to help the inner ear in some way. I also tried hard to follow the recommended low salt diet.[39]

After the MRI, I had no more contact with doctors for that year and let up on efforts to see what could be done to reduce symptoms or get rid of the condition. Since I worked in a school, I had the summer off and after the middle of July, was mainly normal in my energy and functioning during the vacation, despite those occasional reminders of the condition lurking within. Returning to work in September also meant returning to increased symptom levels, but I had no overwhelming episodes of vertigo for several months. The doctor had said a low level of symptoms might last for several months, so I was riding with that thought, waiting it out.

I noticed I was able to believe two completely irreconcilable ideas. On the one hand, I could believe I would never entirely get over MD and that even if I felt well for months it would still be there, hiding in the shadows. At the very same moment, I could also believe that the disease was going to disappear, that a day of feeling well was the first step on the road to full recovery. I knew this was not so likely, based on what I had read about the course of the condition – which was that symptoms of MD were likely to recur over time. But believing I was

---

39   See Chapter 7, "Initial Treatments for Vestibular Illnesses," for an explanation of these.

going to get well anyway, despite all I'd been told, was consistent with my image of myself as a vital person. All my life, I'd loved to be outside doing active things. As a child, I loved horses and riding, running through the woods, and climbing trees. My childhood games involved imagining myself conquering dangers, being free and wild. As I grew up, I continued to like sports and activities that give a sense of motion and sometimes offer excitement – skiing, surfing, swimming, dancing, gymnastics, hiking. Quite apart from the fact that it's a challenge to simply walk around when vertigo strikes, never mind engaging in sports, I found that just knowing it was hanging around somewhere inside my head, even if I was feeling well, eroded my sense of myself. I became somewhat more inclined to see myself as being a weak and vulnerable person, rather than someone eager for adventure.

7

Initial Treatments for Vestibular Illnesses

A treatment is not always the same thing as a cure. Many times, the main point of a treatment is for symptom relief – the idea is to make you more comfortable. When you have a cold, a doctor might prescribe or recommend cough suppressants and decongestants, but you both know you are likely to recover from the cold with or without medicine. So the medicine is used mainly to help you feel better while you live through the cold. It functions solely as a treatment. However, if it serves to stave off a more serious infection, such as bronchitis, it may also function as a preventative. A treatment that actually stops a disease is a cure. This all sounds a bit like nit-picking, but the issue of treatment, partly to reduce the severity of symptoms and but with the strong hope of preventing future occurrences of vertigo (and hence, to cure it) looms large in the minds of many people with vertigo, and sometimes adds to the confusion they feel.

Initial Treatment for Menière's Disease

For an individual diagnosed with MD, the initial treatment recommendations you are likely to encounter are often presented as a short list. Very briefly, drugs to reduce the vertigo and sometimes nausea as well, a prescription for a **diuretic**, and dietary restrictions comprise the list. To minimize the sensation of vertigo, drugs similar to Dramamine, familiar to most as a preventative for seasickness, are usually recommended. Some of the trade names are Meclizine, Antivert, and Bonine. Benzodiazepines such as Valium or Atavan, typically used to address anxiety, are often prescribed since they are thought to quiet the inner ear. Anti-nausea medications that are sometimes prescribed are known as **antiemetics**. Diuretics are often referred to as "water pills," with hydrochlorothiazide or Triamterene being commonly prescribed.[40] To help reduce the intensity and frequency of attacks, or possibly even prevent them, the recommended dietary restrictions include reducing salt intake, and eliminating, or at least cutting back on, caffeine and alcohol.

The thinking behind restricting salt and taking a diuretic has to do with maintaining a more favorable balance of fluid in the inner ear. As noted in chapter 3, endolymphatic hypdrops, or too much fluid in the inner ear, is thought to be causing the symptoms of MD. Since salt increases fluid retention throughout the body, it may also do so within the ear, hence consuming less of it may be helpful. Diuretics are the other side of that: since they cause you to secrete more water, they may also help get rid of that excess fluid. Some people take the diuretics only during an attack, others take them daily in the hope of forestalling attacks. I took them daily for about six years. Steroids, certain vitamins, and sometimes antihistamines may also be prescribed or recommended by your doctor. Avoidance of stress and adequate sleep are usually advised, as well. Generally, none of these things is thought of as a cure – they may help you to live more comfortably with vertigo, but are not likely to eliminate it.

Limiting salt can be confusing, since it permeates most of what we eat. 2000 milligrams a day is the standard nutritional recommendation for everyone, including totally healthy people. Since the average person actually tends to eat more than this, often way more, some doctors or websites will tell individuals with MD just to make sure they don't take

---

40   A diuretic is an agent that makes you urinate more frequently, although I never noticed this happening in my case. Hydrochlorothiazide was originally developed as medication for high blood pressure.

in more than 2000 milligrams. More often they will recommend getting the salt down to 1500, or even 1000 mgs a day. This is not as easy as it sounds. Whenever you eat out, you will be getting giant helpings of salt. When you eat at home, you can read the amount on packaged foods, but will be disheartened to see how much salt is found in ordinary comfort foods like canned soups or in any processed food. A lot of people frequently eat sandwiches made with meats from the deli for lunch, which also tend to be very salty. You can control how much salt goes into the food you cook from scratch, but have to measure very conscientiously. 2000 mg of salt is only one teaspoon! In addition to the nuisance, people crave salt. And like any restriction to your diet, it's hard to give it up. But if you do cut down on salt, you will get used to a less salty flavor over time, and won't crave it as much. You may also find that some herbs and spices make up for the less salty flavor, at least to some extent.

Curbing the use of caffeine is less confusing than cutting back on salt, as long as you are clear about which foods – other than coffee and tea – contain caffeine, such as chocolate and colas. Obviously, those dependent on their morning cup to wake up in the morning won't embrace this restriction with enthusiasm. Cutting out alcohol is likewise a clear directive. The main question that may arise is whether to continue avoiding these substances during times when you are free of vertigo.

Early on, I worked at maintaining a regimen of limited salt, took hydrochlorothiazide daily, and Valium or Meclizine as needed, while waiting to see how things evolved. For years, I had used very little caffeine, so that didn't represent a change for me. I didn't drink alcohol often anyway, and certainly had no temptation to do so if I was the least bit woozy. Since my symptoms bothered me to a varying extent for months, it was never clear whether or not using less salt was making a difference. It certainly wasn't making the kind of difference that aspirin can make for a headache, or an antibiotic for a sinus infection. The level of symptoms came and went in a way that seemed entirely random, nothing to do with what I'd eaten. Seven years down the line, this has remained true for me. In my case, it is not a bit obvious that a rigorous low salt diet reduces symptoms or extends the time between attacks, or that being casual about salt makes matters worse.

It is important to follow this section up by noting that, among my interviewees, the majority of those who were actually diagnosed with MD were advised to follow a low salt diet and to avoid caffeine and alcohol. Many of them felt as I do – either that it made no difference or that it was hard to tell whether it did. There was, however, an enthusiastic group that experienced the regimen as a valuable way to keep symptoms under control. The most dramatic example was that of Kim, who had been suffering intermittently with symptoms for around 30 years. Her symptoms had been increasing in frequency and severity over the ten more recent years, but she had not derived much benefit from consulting with her doctors. She had felt "so disregarded," and was afraid she might need to go on disability, especially since driving was part of her job. She was finally referred to an ENT, who told her about the importance of cutting back on salt, along with caffeine, and gave her a prescription for a diuretic. For her this made a miraculous difference, and she reports feeling "wonderful" by comparison since then. It's remarkable she'd never been told about it before, and great that she could feel so positively about it after years of feeling neglected. If a low salt diet does help you, then you profit from a treatment that has no worrisome side effects and that also gives you some control over the condition.

The use of Meclizine, Antivert, or Bonine was also recommended for the majority of the people I spoke with. About half of them found these to be helpful, and some felt it desperately important to have them available at all times. A couple of people said they wouldn't leave the house without it, no matter how long it had been since they'd had an attack. The other half didn't feel these medications made much difference. Similar comments were made about Valium. Steroids didn't seem to be part of the routine initial therapy for most. A couple of people had been prescribed a Scopolomine Patch, and found it helpful. The patch is normally used to treat motion sickness. Another person said using special wristbands that have been developed to prevent motion sickness worked well for her. These are inexpensive bracelets that exert a slight pressure on the inside of the wrist, and carry no risk of side effects – hence, a great thing if they help you. You can find them in drug stores.

Thus, there are a variety of regimens that may be recommended to a person when first diagnosed with MD, but none of these is expected to stop the disease in its tracks. The main expectation is that they may reduce the intensity or frequency of the symptoms. There are many who feel satisfied with this initial treatment protocol, even if they continue to be afflicted with periodic bouts of vertigo. These people see a direct benefit from the regimen, and thus are motivated to follow it. Among the people I interviewed, they were in the minority. One problem with this standard set of recommendations, either as advised by a physician or found online, is that they tend to be presented as if they were surefire – as if you could expect reliable benefits such as you get from an antibiotic for bronchitis. Little uncertainty is expressed, which is fine for the patient if the protocol is helpful, but if it is not, it only adds to the confusion and anxiety.

The people who saw no connection between the fluctuations of their symptoms and the dietary restrictions, or those whose attacks tended to have a major impact on their lives, as well as those whose lives had been altered – sometimes radically – by vertigo remained interested in finding a *cure* for MD, sometimes desperately so. More about this later.

Transtympanic Steroid Injections

Although I didn't find the injection of steroids into the ear within lists of standard treatments for MD, I include it here because I've come across a few articles recommending it, and because it is not seen as a radical procedure. In an article by Norman T. Berlinger, MD,[41] published in 2011, it is described as providing good control of vertigo,

---

41   Berlinger, Norman; "Meniere's Disease, New Concepts, New Treatments" Minnesota Medicine, 2011 November; 94(11)

for those who were not helped by the usual regimen. The article also referred to it as a "new and powerful" therapy. This surprised me, since references have been made to the use of steroids for many years. Be that as it may, such injections are not endorsed by all, as is true of many approaches to treating MD. The Mayo Clinic website simply presents it as one of many choices, without seeming to endorse it. However, Augusto Pietro Casani, MD,[42] conducted a survey, comparing the outcomes of gentamicin (an antibiotic) and dexamethasone (a steroid) injections, and found that gentamicin achieved far better results than did dexamethasone. 81% obtained "complete control" with the "gent," while only 43% did with the steroid. With the more recently developed slow doses of gentamicin, he found a low incidence of damage to hearing. On the dizzy-and-balance website, Timothy Hain MD,[43] was more actively negative about steroids, seeing it as a "near last resort." He didn't believe the benefits of steroid injections would last, and thought repeated injections over time would mean increased risk of infection and other injury. He cited Casani's work.

When steroids are used, as with the other primary treatments, they are not likely to be considered a "cure." And as noted previously, it is hard to tell whether a given treatment is making a difference because of the natural history of vertigo. Even when attacks feel unbearably long, lasting for a few days, in most cases they still begin to ebb away. When they recur, they don't usually follow a pattern, so if you take a course of steroids followed by a long break, you can't tell whether the break would have happened anyway, or whether you have the steroids to thank. I had two courses of oral steroids when the condition first began, and both times I took them when my symptoms had begun to lose intensity. Thus, I didn't really credit them with my (temporary) recovery. During my second year, I had two transtympanic injections of a steroid, which was then followed by several months of only minimal symptoms. Maybe they did help – but I never felt sure about it.

---

42   Casani, AP et al; "Intratympanice Treatment of Intractable Unilateral Meniere's Diseaase: Gentamicin or Dexamethasone?" <u>Otolaryngology Head Neck Surgery</u>, 2012 March 146(3)

43   www.dizziness-and-balance.com/treatment/it-steroids.htm Page modified Jan 13, 2013

In considering steroids, Jake, whose experience with surgery will be described later on, found that as a result of intensive steroid therapy for a nasty case of poison ivy, all of his lingering (and "tolerable") symptoms of vertigo vanished, and gradually returned after the medication was terminated. This was particularly unexpected since a short course of oral steroids soon after the onset of MD had not helped him. So – here was a mysterious benefit! It may be that the first course of steroids was simply not strong enough to help Jake. In any case, the more intensive course did not provide a lasting cure.

Initial Treatment for Benign Positional Vertigo

Unlike the situation with MD, people suffering with BPPV *do* have a treatment that can clearly put a stop to a given attack of vertigo, and in many cases will even effect a permanent, or semi-permanent cure. The treatment was briefly mentioned in Chapter Three, and is known by the name of its inventor: the Epley Maneuver, for John Epley. He first presented it to a medical conference in 1980, where it was unenthusiastically received. In 1983, he tried to submit a paper about it to several journals, with no success. The maneuver did not receive general acceptance until late in the 20th century, nearly 20 years after Dr. Epley's first attempt to publicize it.[44]

Luckily, the majority of people presenting with symptoms of vertigo are found to have BPPV, rather than MD. The basic symptom is vertigo itself, along with the accompanying nausea or queasiness and the difficulties with balance. Deafness, tinnitus, and head pressure are not included as part of the diagnosis. As detailed earlier, some may be hit by a brief sickening whirl that leaves within minutes, a disconcerting event that sometimes happens a few times a day. Others will have extended periods of vertigo similar to those experienced in MD. Most of the time, especially with the brief episodes, the vertigo is precipitated by specific motions of the head or body, which is not generally true of MD. Unlike MD, some people with BPPV develop the symptoms

---

44   More about this interesting story can be found in Chapter 20, "History."

following some sort of trauma to the head – not always immediately after, and not necessarily anything dramatic like a concussion or skull fracture. In such cases, there is less sense of mystery regarding causality, since it makes sense that things might get knocked out of place when you bang your head. Remember that, in BPPV, the tiny crystals leave their nests in the vestibule to float around where they don't belong – in the semicircular canals. But much of the time no one has any idea what makes the crystals break free.

The trick in treating BPPV is to maneuver the crystals back where they came from, or at least to shift them out of way. The amazing thing here is that the "Epley Maneuver" is an effective way to do this, and it involves no medication and no specialized equipment. In a nutshell, it is done something like this: The patient is seated on a bed or special table, and turns her head toward the affected side, then lies back. After a pause, she turns her head toward the "good" ear. This is followed by rolling onto that side, then by rising to a sitting position.[45] Usually, increased vertigo is experienced during the procedure, and possibly for a while afterward, but in the majority of cases, this will subside soon, and that attack is over. Despite the wonderful benefit, the patient should be prepared for the possibility that undergoing the Epley can be an overwhelming and sickening experience. Some people may be able to do the procedure for themselves by following instructions given them by the doctor or physical therapist, or found on the Internet.[46] However, most will need an experienced practitioner at least at first, particularly if they feel worse while it's being done. Many doctors know how to do it, but not all, and some may not even be familiar with it. An ear, nose, and throat specialist should know how to do the maneuver, even if a regular health care provider or the emergency room does not. Many physical therapists have also been trained to do it. In some cases (but again not all), the maneuver will be tried the first time you contact a medical facility with the complaint of vertigo. Although the maneuver

---

45   These are not directions on how to do it! It's presented just to give a general idea.

46   Videos showing how to do the maneuver, as well as detailed descriptions of it, can be found on the internet by googling "Epley Maneuver."

typically works right away, some of my subjects reported needing a follow up treatment a day or so later, sometimes more than that.

It should be noted here, that if a patient is lucky enough to be evaluated by a medical practitioner who is familiar with the Epley, part of your evaluation might include the Dix-Hallpike Maneuver, again an eponymous label. This proceeds somewhat like the Epley, but is shorter and may be done to determine whether the Epley is likely to work, or to figure out which ear is the one that's involved. Like the Epley, it will stimulate nystagmus, which will beat faster in the direction of the affected ear. If the Dix-Hallpike is performed after the Epley Maneuver is completed, and no nystagmus occurs, this is a strong indication that the Epley was successful.[47] Some practitioners recommend that the patient sleep sitting up, or with the head elevated for a night or two after the procedure, to help stabilize the placement of the canaliths. In trolling through the Internet, I found that there was variation in whether or not this was thought to be necessary.

For the majority of people with BPPV, the crystals are loose in the posterior canal, or occasionally in the anterior[48] canal. In a small number of cases, it's the horizontal canal where the trouble is. In that situation, the Epley is less likely to work and a maneuver called the "barbeque roll," or Lempert Maneuver, may be used instead. This involves primarily a specific procedure for rolling the head.[49] Being less common, it is unfortunately likely to contribute to delay in finding effective treatment, since fewer professionals may be familiar with it.

Consider the story of Harold, a man in his 60s, who was pushing a shopping cart through the aisles of a large grocery store several years ago, when he began to feel dizzy and uncertain of his balance. He held tightly to the shopping cart to steady himself. Feeling ill and somewhat nauseated, he was very worried that something serious might be

---

47 There are several other variations on the Epley and Dix-Hallpike Maneuvers, named for the people who developed them. They all involve movements of the body designed to reposition the canaliths. The following are the names of those I have come across in my reading: Semont Maneuver, Brandt-Daroff Maneuver, Li Repositioning Maneuvers. Another variation is called the Canalith Repositioning Procedure (CRP) obviously not named for a particular person.
48 Sometimes referred to as the "superior" canal.
49 Another eponymously named maneuver for these circumstances is the Appiani.

wrong. His wife, who was with him, helped him to find a doctor quickly. Luckily, the doctor turned out to be someone skilled in the Epley Maneuver, so after she took a history from Harold she administered the maneuver. (In this case, no Dix-Hallpike was done.) He walked out the door of her office feeling fine and very relieved. He never received any diagnosis and the dizziness has not returned. Technically he may not warrant being labeled BPPV because the situation doesn't seem to be recurrent, but it's very likely that a few of the canaliths had somehow moved out of place. One of the benefits of the Epley here, apart from bringing the symptoms to a halt, is that when it works it indicates that the problem lies in the inner ear, rather than the brain. Thus, concerns about strokes, brain tumors, or random neurological disorders can be put aside.

Narratives of People with BPPV

Stephanie reports an initial bout with vertigo five years ago at age 70, waking up in the morning and being unable to get out of bed, looking at the ceiling rotating above her, feeling scared. The vertigo lasted several hours, while feeling off balance, queasy, and having funny feelings in her head lasted a full day. She herself is a doctor, and immediately called a neurologist, who met with her promptly and ordered an MRI. The MRI was normal, and he told her the symptoms were benign but probably would recur. For the next four years, she had occasional mild episodes, but nothing that really knocked her out of commission. Then in the fall of 2010, she got out of bed and fell. She felt extremely dizzy although still marginally able to function, walking through the house holding on to walls and furniture. When she looked up, the dizziness would wash over her. When she lay down, it seemed worse than when she was standing. She waited a few days for it to subside, but when it persisted, she made an appointment with an ENT. He performed the Epley Maneuver, which she found frightening due to the intense vertigo it generated, followed by relief when the vertigo disappeared. He gave her the diagnosis of

BPPV, and advised her to sleep upright for a few days to help the canaliths to settle in place. Over the subsequent years she had no further attacks, felt a greater understanding of the condition, and is no longer frightened by it. She has completely stopped drinking alcohol, which she says is not logical, but helps her feel she is doing something to combat the attacks. Clearly, Stephanie profited from the Epley Maneuver itself, as well as from a sense of understanding.

Natalie is a woman who has a diagnosis of BPPV, characterized by strong and recurrent vertigo. She first experienced an episode of vertigo in her twenties, while living alone. She simply woke up one morning and found the room to be spinning. She discovered that she could reduce the sensation by lying on one side, but was unable to get out of bed or walk without support. Because she works as a nurse, she was pretty sure of what was happening and called her doctor, who wrote a prescription for Meclizine. A friend brought the medication over to her, and – lo and behold – it worked, and the spinning subsided. Natalie's life went on without another episode for ten years or so, when she began having much more frequent attacks – about three a year, some very severe, lasting for as long as three or four days. Around this time in her life, probably in early her 40s, a doctor told her over the phone about the Epley Maneuver. Based on his description, she tried doing it on her own, but it made matters worse to the extent that she was unable to function for several days, and spent most of the time sleeping. Some time later, she learned of a physical therapist trained in the technique, and has since found the maneuver works well for her when it is done right. She notes that the treatment itself can be "horrifying" because of the degree to which it temporarily intensifies the spinning.

Knowing that something can be done to stop an attack has made a huge difference in Natalie's life. However, she has continued to live with the threat of an attack coming on without warning. She remembers a time when she was away from home, staying in a hotel with her husband, when vertigo struck. In addition to the extraordinarily sickening feeling of the whirling room, she felt trapped. She needed to get to the physical therapist, who was not close at hand, but going anywhere seemed impossible to her. Even though her husband was there to drive her home she could not imagine tolerating the long car ride, or even riding down in the elevator in order to get to their car. She did have Meclizine with her and found that it took the edge off and enabled her to leave the hotel, leaning on her husband for support. He drove her to the therapist for the Epley, and she was able to walk out of that office feeling well, as is usually the case for her. In a limited sense, Natalie is lucky in that there is a specific treatment for her attacks, and one that does not require drugs and their attendant side effects. However, because no treatment has yet been found that can avert the vertigo altogether, and because it is unpredictable, she lives with apprehension and has sometimes avoided making plans to go away with friends for fear of suffering an episode and ruining everyone's good time. Natalie's case is more complicated than many of those with BPPV, since vertigo is not her only symptom. She has a hearing loss and tinnitus as well, but does not believe these have anything to do with the vertigo.

Neither Stephanie nor Natalie have the brief episodes of vertigo that characterize BPPV in the classic definition, nor do several of the others whom I interviewed. However, they must all have otoconia that have moved out of place, which is the underlying supposition for BPPV. I've tended to assume that whenever the Epley Maneuver is effective it signifies that the patient has BPPV, perhaps a kind of backward reasoning. But physicians do not expect it to be helpful for those with MD, since it would not eliminate endolymphatic hydrops, nor would they expect it to work for an infection like vestibular neuritis or labyrinthitis.

To sum up, in my sample most people diagnosed with BPPV who were offered the Epley Maneuver found it be helpful, and in some cases, curative. It remains amazing to me that the Epley is still not used everywhere, and that it or the Dix-Hallpike Maneuver are not part of a routine evaluation procedure when someone with vertigo first sees a doctor.

Initial Treatment for Labyrinthitis

As noted earlier, labyrinthitis is an acute infection, usually caused by a virus. It is not life threatening and is expected to resolve on its own. Thus, in most cases, the initial treatment for labyrinthitis stands as the only treatment that will be administered for the illness. As with other vestibular illnesses, that treatment will consist of medication to diminish vertigo, such as Meclizine and Antivert, as well as an anti-emetic to help control nausea and vomiting. If the condition is triggered by a common middle ear infection, which is usually bacterial in nature, then an antibiotic may be prescribed. If the infection is not bacterial, some would recommend a course of steroids or antiviral medication. Especially if recovery is slow, vestibular rehabilitation therapy (a sub-specialty of physical therapy) and exercise are both recommended in order to facilitate adaptation to the assault on the balance system.

For a small number, labyrinthitis in a mild form may become a chronic condition, with some imbalance and a low level of vertigo continuing on. Again, vestibular rehabilitation therapy can be effective in dealing with this, by "retraining" the brain to respond differently to the signals from the inner ear. But generally speaking, the prognosis for labyrinthitis is for a full recovery.

Initial Treatment for Vestibular Neuritis

The initial approach to treating VN is also focused on helping the patient to feel more comfortable. Just as with MD or labyrinthitis, you may be given medication to reduce vertigo or an anti-emetic to control nausea. You might also receive a prescription for Valium or Klonopin to help quiet the inner ear, or possibly antihistamines. Since it is believed that VN is caused by a viral infection, the greater likelihood is that you will recover spontaneously from the attack, just as you are likely to do from most other infections. Some physicians think that a course of oral steroids accelerates recovery. In addition, some feel that an anti-viral medication, such as acyclovir or Valtrex, makes sense, since the illness is believed to be caused by a virus. Engaging in as much normal activity as can be tolerated is thought to aid recovery, too. This can help your vestibular functions adapt more quickly and get back to normal.

Living through the vertigo caused by VN can be a grueling experience, but if the patient is lucky, it will end up being just another illness experience, meaning that you feel terrible for a few days or weeks but then recover. When it's over your life gets back to the way it was beforehand. However, there are times when VN does some actual damage to your inner ear or the vestibular nerve. If this happens, you may continue to experience some imbalance or dizziness that drags on. In this situation, vestibular physical therapy (discussed below) can be extremely helpful.

Louisa is one of only two people with whom I spoke who was diagnosed with VN. Symptoms of vertigo began for her in her late 30s, 17 years before we spoke. Her diagnosis at that time was BPPV, the kind characterized by intense symptoms, like "falling off the Empire State Building," "violent jerking horrible movement." As long as she remained still she was okay, but any movement precipitated the vertigo. She learned to move slowly and with great care, and was affected for nine months. She then felt reasonably well, suffering occasional mostly minor episodes of vertigo during the course of the next several years.

Four years before we talked, when in her early 50s, Louisa woke up in the night with another terrible attack, worse than any before. Again she felt herself to be dropped from a huge building, plummeting through the air. When she opened her eyes she saw "objects in duplicate all spinning." Her heart was pounding, she was sweaty, breathing hard, terrified, vomiting. She crawled across the room to find her cell phone, but couldn't read the numbers because they were all in multiples. She did finally reach a friend, who came over and called an ambulance for her. She blacked out as the emergency medical technicians got her on the gurney and started IV Compazine to help control vomiting. She stayed in the hospital for a few days, during which time she had trouble getting out of bed and needed a walker to get around safely. Her muscles didn't seem to work well, her depth perception was off, and she still saw the world around her in multiple images. The vertigo subsided but didn't disappear. Despite all this, she was discharged with a diagnosis of BPPV. She required first a walker and then a cane to support her balance for several weeks. She was unable to drive for a while so friends pitched in to help her get around.

As an outpatient, she was referred to an otologist at a large university hospital who told her she had VN and recommended vestibular physical therapy. This proved to be very helpful to her, but her recovery was gradual. The experience of pronounced multiple vision lasted for four months, and she was too impaired to do much of anything for about three months. She sat in a chair all day, unable to read, watch TV, or knit. She heard "chimes" in her head that "took forever to go away." She worked with the PT over several months and began to resume normal life, but didn't really feel confident that the nightmare was over until around a year ago. When she wakes up in the morning now she still looks around the room to make sure there are no extra images anywhere.

The above is a sketchy summary of Louisa's complicated and remarkable encounter with vertigo. Early on, her symptoms were clearly set off by moving around, and the diagnosis of BPPV may have made sense at that time. Her experience was definitely dramatic, but was unlikely to have been attributed to MD since it had no impact on her hearing. She has some residual problems with balance: she can't walk steadily in the dark. The fact that PT made such a remarkable difference for her is heartwarming and encouraging. In several articles about VN that went on to discuss how to address the situation if the individual fails to make a good recovery, vestibular physical therapy was often recommended.

Vestibular Physical Therapy

Physical therapy is not necessarily a first line treatment for symptoms of vertigo, nor is it usually expected to be a cure. It is, however, often very helpful, particularly if the physical therapist (PT) is trained in vestibular disorders, and thus offers vestibular rehabilitation. In that case, the practitioner will be a specialist who knows a lot about the inner ear, and takes an interest in working with the client to minimize the intrusion of symptoms into daily life. Not every ENT will make a referral to PT, so it may turn out that patients will need to inquire about the possibility or seek it out on their own. As noted above, Natalie always went to a physical therapist for the Epley Maneuver, but overall very few of the people I interviewed had experience with it, even though PT can be valuable in many other ways.

About a year after her first episode of vertigo, Helen was referred to a PT by her internist. She was lucky enough to live in an area where there was a clinic specializing in vestibular problems. She'd had two vertigo attacks during that year, both successfully treated by the Epley Maneuver. When she met with the PT, in addition to working with the Epley, she learned some "habituation exercises," to help her eyes, brain, and inner ear work together. Although she was not symptomatic at the time she met with the PT, she found it calming and helpful. She was treated with sympathetic interest, was able to get questions answered, and the exercises gave her a sense of being able to do something active to combat the vertigo.

Michaela was referred by a neurologist to a physical therapist for the Epley about two years after her initial attack, which took place in 2003 when she was 55. Like many of the others, her diagnosis may have seemed unclear. She was aware of some hearing loss as well as tinnitus, and an unusual doubling of voice sounds when singing. During her first attack, she experienced "dizziness," characterized by the room "scrolling up and down," and an inability to walk independently, which lasted about an hour. She was frightened and made an appointment with her primary care doctor right away. He prescribed Meclizine, which did help, and ordered an MRI to rule out anything ominous, and told her that probably the crystals in the inner ear had moved out of place. Over the next two years, she experienced four more such incidents. Nearly two years subsequently, she had an especially intense attack while away from home. She felt "weird" while going up stairs, and when she got in her car was able to drive only a block before needing to pull over. Again, the environment was scrolling. In particular, she was completely unable to read street signs because of this. She experienced overpowering vomiting, after which she napped in the car until she felt able drive home. This time, she consulted with an ENT, who referred her to the neurologist who referred her on for a consult with a PT. The Epley Maneuver was performed, which proved to make a big difference to Michaela. In addition, the PT offered some other exercises to help with gaze stabilization and control. She used to feel the attacks were "ruling" her life, but now she feels much more in charge of herself. She does wish she'd known of the Epley sooner.

After speaking with Michaela, I communicated with a physical therapist to learn more about what PT could offer. Richard Purdy, a practitioner who has specific training in vestibular rehabilitation, was kind enough to answer a series of questions I sent to him.[50] Most commonly in cases of BPPV, the canaliths in the inner ear that have moved out of place are floating in the anterior or posterior canals of the semicircular canals. As noted earlier, this situation is usually helped by maneuvers that reposition the canaliths, of which the Epley Maneuver is one example. However, there are times when they move into the horizontal canal, or get stuck somewhere, thus requiring other maneuvers. Practitioners trained in vestibular PT are particularly skillful at figuring out what is needed. Purdy does teach patients how to perform the appropriate maneuvers at home, but says it is important to be sure it is done correctly. Through vestibular PT, people with BPPV can also learn to manage their vertigo in such a way that the feelings are significantly reduced or even eliminated.

Vestibular PT uses several techniques that can be of great help not only to people with BPPV, but also to those diagnosed with other vestibular disorders, such as acoustic neuromas, vestibular labyrinthitis, neuropathy, trauma to the inner ear, or head trauma. It can be helpful to people with MD between episodes if they have a loss of vestibular function and find that their symptoms are triggered by motion. With the help of PT, people afflicted with vertigo, whether it's of vestibular

---

50   A more detailed synopsis of my questions and his answers can be found in the back of this book.

or "central" origin,[51] can improve visual stability with head movement, decrease motion sensitivity, and improve their balance. Those whose symptoms are apt to be kicked off by hyper-stimulating environments, such as supermarket aisles, loud sound, or overpowering smells can learn to control their reactions better.

For patients interested in consulting with a PT trained in addressing vestibular disorders, it may require some sleuthing to find one. Since few people mentioned working with a PT, it seems likely that referrals aren't made frequently. And because it is a sub-specialty, there may be lots of areas of the country where no such physical therapy is available.[52]

Moderation of Stress

For both BPPV and MD, apart from alterations to diet, medication, or the use of the Epley Maneuver, there is usually the recommendation that stress be avoided, along with getting enough sleep. Theoretically, it should be easy to be conscientious about getting a good night's sleep, although we know millions of people don't do it. There is plenty of evidence that the advent of good interior lighting, especially with electricity, has led to an enormous decrease in the number of hours people sleep. But since fatigue is so often a companion to, or the aftermath of, a bout of vertigo it certainly makes sense to get in bed an hour or so earlier than usual, especially if you typically sleep less than seven hours.

Regulating stress is much more complicated. The majority of people probably consider their lives to be stressful, but that doesn't mean it's obvious what you can do about it. Juggling the responsibilities of work and home, scheduling myriad things to do each day, meeting deadlines, worrying about finances, being in a hurry, dealing with the demands of other people - all are part of every day life.

Vertigo itself is stressful, of course. It generates enough anxiety in

---

51 Remember "central" means it arises within the brain instead of the vestibule. The traumas and acoustic neuroma mentioned here would be central.

52 Lists of PTs may be obtained from VEDA (Vestibular Disorders Association.) According to Richard Purdy, Emory University in Atlanta can provide a list of practitioners who have passed an exam in vestibular rehabilitation.

some people to trigger a pounding heart, sweating, and rapid breathing. It may jack up blood pressure. The autonomic nervous system is releasing adrenaline into your blood stream. Intense anxiety, experienced physically in this way, can be nearly as disturbing as vertigo itself, and may also give rise to frightening ideas about what's wrong, or what's apt to happen next. Experiencing a "nightmare carnival ride," Kim thought this was how a schizophrenic must feel, with voices inside the head and a deep sense of confusion. Many people at first assumed they had a mortal disease, or might be dying, and a few *did* think they were going crazy. Having a stroke or a brain tumor were fears that crossed the minds of many, also. Even learning that you are not suffering from a terminal illness or going crazy does not necessarily prevent anxiety while in the throes of an attack. The experience itself is so distressing that it stirs up the autonomic system anyway.

Whether or not a person is subject to acute anxiety during an attack, there is almost nobody who doesn't have some level of anxiety about the possibility of a recurrence between attacks. When special events, holidays, and vacations are planned, the fear that you may be struck with vertigo and unable to participate looms in your mind on some level. Especially if attacks are frequent or particularly intense, there is concern about the effects the condition may have on your whole life, never mind holidays. Many people are deeply troubled by this sense of uncertainty. So ironically, people subject to vertigo receive an extra helping of stress, at the same time they are advised to reduce it. And those who live with stress every day most likely live with an elevated level of "stress" hormones (cortisol and adrenaline).

Taking one day at a time, concentrating on one task at a time, and finishing what you start before moving on, are all generally helpful techniques for pacing yourself. Regular exercise can make a difference, although it's not always possible when your head is swimming and you are actively off balance. But in between bouts, exercise can have the effect of lowering blood pressure, lowering heart rate, and counteracting the effects of stress physiologically.

Learning to use deep slow breathing techniques is very valuable, need not take up much time, and is worth doing regularly. Like exercise, it can relax your whole being. Physiologically, deep breathing and relaxation slow the heart and breathing rates, actually counteracting the effects of the "stress" hormones. Thus, despite their simplicity they are a *bona fide* treatment. It is worthwhile to spend ten or fifteen minutes each day, seated in a comfortable chair with closed eyes, breathing in deeply through your nose, holding your breath for a few seconds before letting it out slowly through your mouth. While doing this, allow your muscles to relax so that you feel like a rag doll. Learning meditation and/or mindfulness is a bigger project, best done through taking classes, that helps to quiet the mind, alter some biological rhythms, and diminish anxiety.

Another factor that stokes up stress for people with repeated bouts of vertigo is the fact that many have felt frustrated, or sometimes humiliated, in their search for help, perceiving themselves to be in a situation of little interest to physicians, and for which there is no hope. This will be explored later in chapter 16.

Yoga, Tai-Chi

Like meditation and deep breathing and even physical therapy, yoga and tai-chi are not necessarily going to be suggested to you by your doctor. Yet they are all practices that may help you to feel much better, and are not likely to be medically contra-indicated. The famous Mayo Clinic[53] does recommend tai-chi for stress reduction, describing it as "meditation in motion." It is a Chinese art, consisting of a series of slow movements and positions that help with balance and focus. With the disclaimer that people should check with their doctor about its suitability for them, it is usually something people of any age can do. Yoga is an Indian discipline, and is probably more physically challenging than is tai-chi. Deep breathing and focus are also part of it. One person with whom I spoke, when in the throes of distress about the effects of vertigo on her life, said that taking a yoga class helped her to feel human again. Classes in either are widely available.

---

53   www.mayoclinic.com/health/tai-chi/

Addressing Tinnitus

There does not seem to be much specific treatment for tinnitus yet. However, a device has been developed that can be worn over the ears to mask or reduce the sound. No one I interviewed mentioned trying these. I found one recent study that indicated that a form of Cognitive Behavior Therapy was effective in addressing tinnitus. The study had a good number of subjects: 492, half of whom were controls and half of whom received the therapy. Even those with severe tinnitus benefited from the treatment.[54]

---

54   See Bibliography for Cima, Rilana.

8

More about the Experience of Vertigo

How vertigo is experienced varies. In meeting with doctors myself, I never got the impression that the way I experienced it was thought of as having any special significance, nor have I found any suggestion in my readings or talks with people that other professionals think of the variations as mattering. However, it strikes me as a possibility that they may be indicative of different underlying processes.

In addition to what was described in Chapter Two, another relatively common, but not universal, experience associated with vertigo is the feeling of falling. It can be deeply disorienting. Some people report that you "cannot place yourself in space, can't tell up from down." People lying on their beds may cling to the frame, feeling sure that it is tipping and they are about to roll off. Or they may be seated on a chair and feel that it or the room is tilting, and grab for the nearest stable thing. No matter what they tell themselves, the sensation of falling is stronger than the rational thought ("I know I'm safe on my regular old bed") and they hang on for dear life. On at least one occasion, I felt as if my bed was dropping straight down, as if on a freight elevator whose cables have snapped, which sounds as if it might be fun, but it was not. I clung to the headboard. Also in my own case, I have been woken up in the night by a repetitive dropping sensation within the left side of my head, as if a small roller coaster were zipping down into the trough, then back again, then down again, back up again.

A certain number of people seem particularly prone to that sense of free-fall, described by Yvonne as feeling "the earth was opening up to swallow me." Richard spoke of a "drop down an elevator shaft," and Louisa found the world was "crashing" – she was being violently thrown into space, "like falling off the Empire State Building." It is like "a nightmare carnival ride." Marnie noted that it seemed that a "big creature throws you back on the bed." It's safe to say it's an experience accompanied by fear, the sense of unstoppable falling feels so real. It interests me that no one reported feeling that they were lifting off from the earth and flying – it was always a downward plunge. It's important to underline that these experiences are not exactly the same as feeling that you're about to fall when you can't maintain your balance. Sometimes they last longer, for one thing, and may be repetitive – you may be dropping off the skyscraper, then inexplicably be back up at the top doing it all over again. However, people may fall or hurt themselves as a result of things they do to stave off the sensation, like grabbing hold of something unstable (a lightweight chair, a stack of books) in order to steady themselves.

By its very nature, vertigo changes one's relationship to the world around – the familiar becomes strange and unpredictable. It alters your perception of a world that seemed familiar and stable just a few moments before. Esther describes it as "almost other-worldly," as if being hurled into an alternate universe. Everyone feels enormously out of control – you are overtaken by a force you cannot stop, and are suddenly unable to perform the most minimal action. A variant of the whirling sensation is to see objects in the room sliding past, somehow getting back in place before they slide past you again. Some people see vertical "scrolling." They look at a street sign but can't read it because it keeps rolling down, then down again, then down yet again. Vision may bounce, go every which way. There's a word for those phenomena: "**oscillopsia**." One's eyes may seem to be uncontrollably rolling around in a wild and crazy way, an experience that feels different from that of nystagmus. Some report having seen objects around them break into pieces "like a kalei-doscope." I have woken up repeatedly with double vision, while a few others talk of double vision, or even vision in multiples, as being part of their most intense episodes.[55]

Other less common and strange phenomena can occur. People prone to vertigo may find that it's triggered by somewhat surprising stimuli. Being in a supermarket, with its enormous array of things for sale, the endless mind-boggling choices, can do it. The cleaning products and detergent aisle has more intense odors than any other part of the store, and some find it the place most likely to precipitate vertigo. Other people find that loud sounds can get an attack started, and – this being hard to understand - not always right away. They may go to a concert and then feel terrible the following day. Two people found they could continue to enjoy concerts as long as they muffled the noise level by wearing earplugs or earmuffs.

A number of people found that being in an environment where things were rushing past gave rise to vertigo, the most disturbing example of this being a highway. Nick found that, after stopping for a light, he would tend to get a burst of vertigo when the cars around him started up again. He had to give up fly-fishing, an activity he enjoyed enormously, because he couldn't tolerate the steady stream of water flowing past him.

During a vertigo attack, there's not much anyone can do but wait for it to pass,[56] other than the use of medication to address motion sickness, nausea, or anxiety if you find them to be helpful. Usually people are more comfortable with their eyes closed, since it reduces the sensation a bit. There's no real comfort to be found until it's over, nothing akin to the coziness of blankets or a cup of tea when you have the flu or a nasty cold. Most people need to sleep after an attack is over, but sleep does not always provide a sense of peace or a "quiet rest when the long day's over." For those of us subject to vertigo, whether it be

---

55   These latter phenomena may not originate in the vestibular system, more likely in the brain. But from the point of view of the person with vertigo, they are part and parcel of the whole experience.

56   However, if the vomiting or diarrhea is extreme, you may become dehydrated, necessitating a trip to the ER.

MD or BPPV, the very act of lying down can precipitate a whirling head or a whirling room – not necessarily very intense or long-lasting, but nonetheless unsettling. Changing position in bed can do the same thing. Even without that, passing symptoms experienced during the day or week can create a low-grade anxiety, leading to an alertness that interferes with falling asleep readily, or prevents getting back to sleep if sleep is interrupted. Getting under the covers may no longer feel comforting, but dreaded, with the anticipation of restlessness throughout the night, or disturbing dreams. As quiet settles through the house, tinnitus sounds loudly in the ear. I remember, during the first year I had vertigo, waking up in the night and thinking I heard the murmur of voices in another room, or perhaps a radio playing. Even though I was sure this could not be the case, I got out of bed to investigate. As I snuck through the darkened house, it seemed alien, and I was alert to the hum of the refrigerator in the background. There was nothing to be seen except the shadowy furniture. I did wonder if I was actually hallucinating, and felt a twinge of horror. The night, filled as it was with mysterious rustling sounds, seemed to whisper menacingly.

That sense of being out of control extended for some people to a fear they were losing their minds, or going crazy. As noted before, many thought at first they might be dying, or assumed they might have a brain tumor or some other horrible disease. Even without such thoughts, the experience in itself is deeply anxiety provoking. Many live in dread of its happening again. Esther, who had not had an attack for many years, never left the house without being afraid an attack would strike while she was away, and thus always carried Meclizine in her pocket. It is not just fear of the dangerous things you imagine could result from an attack, such as driving your car off the road or falling downstairs. Nor is it limited to the fear of the disappointing things that could happen, such as being unable to attend a wedding or celebrate a holiday. It is fear of the sensation itself, which tears you loose from your normal ties to daily life.

In most cases, vertigo hits without warning, although it is very common for people to wake up in the morning with it. They go to bed feeling fine and open their eyes to see the world shaken off its moorings. However, some people do get a little advance notice. Brent reported feeling as if a curtain was dropping in his mind about five to ten minutes before vertigo would begin, enabling him to get somewhere safe ahead of time. More recently for me, I tend to feel dragged out and unsteady on my feet for about a day before symptoms arrive. But like everything else about vertigo, whether or not you get a warning is likely to vary over time.

It is noteworthy that people refer to an episode of vertigo as an "attack." You "come down" with most illnesses – the flu, pneumonia, the mumps, what have you. You "feel," "have," or "get" various sorts of pain – cramps, headaches, even emotional pain such as depression. Use of the word "attack" is consistent with the feeling that vertigo is inflicted by an external force, one that has taken control of you and rendered you virtually helpless. The fact that it alters one's ordinary perception of the world adds enormously to the distress. The experience is so foreign, so removed from almost any other in life. The way you look at the world, the way you count on perceiving the world, is not there. Strange things are happening. Chaos reigns.

9

Disequilibrium, or Balance Problems

Balance is obviously a key concern for anyone prone to vertigo. In our ordinary daily life, it is something we usually take very much for granted. It operates almost entirely on an unconscious level, with your vestibular system working to make sure you are oriented to the space you inhabit - that you can sit, stand, and move about without falling over. Normally, people don't think about it much, although they are aware of situations in which balance may be challenged, maybe during various athletic maneuvers, standing on one leg, or walking aboard a rolling ship. If we stop to think about it, we can see that a baby's sense of balance needs time to develop, unlike most other mammals. Around age one, they stagger a few steps before they plop down on their bottoms. You watch a beautiful progression as they begin to walk with ease, and then maintain their equilibrium as they start running and jumping, flying around the playground fluidly and with confidence by the time they are four or five. In addition to the vestibular system, a good portion of this progression is of course due to the development of muscular strength and coordination, both of which have an intimate association with balance.

We are aware that balance can be improved by practice, and that some people seem clumsier and less well balanced than others, but we don't generally imagine, or develop a fear of, a life without it. From quite an early age, however, we do imagine the possibility of life without sight or sound. People have ordinary experiences with a sightless world, feeling their way around in the dark of night, or snuggling down under heavy blankets in winter. Games involving blindfolds such as "Pin the Tail on the Donkey" or "Blind Man's Bluff" play with the idea. Most folks have some familiarity with a person who is blind or has limited sight, and everyone must know several people who are, if not actually deaf, at least hard of hearing. "Blind" and "deaf" are familiar words. Yet balance is as important to our functioning as are sight and hearing, so much so that some have labeled it a "sixth sense." However, it is not typically seen as a "sense", nor is it accorded the same importance. Perhaps this is partly because a complete and permanent lack of balance is extremely rare, and also because we do not associate it with visible and tangible organs like eyes or ears. Loss of balance is generally thought of as temporary and circumstantial, and even with a vestibular disorder, that is largely true. In fact, the organ for balance was so hidden from us that it was not even identified until the early 19th century.

An interesting thought about balance as a sense (if it is that) is that it is much *more* important to our functioning than is the well-recognized sense of smell – or taste, for that matter. Yet we have much more conscious awareness of smelling and tasting, which we think and talk about every day, than we do balance. We don't think about balance all that much, unless we are trying to learn a new skill that challenges it, or unless there is something wrong with it. Mostly it's in the background, just part of the way our bodies work. But without it, we'd all be lying on the ground like jellyfish, or marionettes with the strings gone slack.

The complexity of the systems that control balance in the body is astounding. While the vestibular system is the center of balance, it is a team player that works in concert with other senses. Previously mentioned were the proprioceptive sensors that operate within your body, sending and receiving information concerning the positions and actions of all your muscles and joints. This is the function of specialized nerve fibers found in all your muscles, tendons, and ligaments, interacting with your vestibular system to keep you upright and moving around effortlessly. They respond to the pressure of your feet on the floor (or tightrope) in order to regulate balance. Feet are where the greatest number of proprioceptive nerves is found, so that people who develop nerve damage there, such as in diabetic neuropathy, have to struggle with balance while walking. People in the throes of severe vertigo sometimes find that moving any part of the body, even just a finger, can increase the intensity of the whirling, emphasizing how interconnected it all is.

Another part in the body that has an effect on balance is the strength of the muscular system. You "lose" your balance if someone lurches into you when you don't expect it, or when you trip. Usually it is quickly regained – your vestibular system snaps into action and your muscles bring you back on track. For most, this will happen without much conscious awareness. But when the vestibular system isn't working well, your muscles have to compensate by working harder, and this often does require conscious effort. Stronger muscles will improve your equilibrium even when there is nothing wrong with the vestibular system. Some part of achieving the wonderful balance of a dancer, a gymnast, or tightrope walker must have to do with the development of strength. Elderly people often struggle to maintain balance when they walk, which may be partly due to weak musculature or diminished proprioception, or partly due to an inner ear that just doesn't work as well as it once did, even if they have no real vertigo. Just as with hearing, sight, and agility, balance tends to deteriorate with age.

Anyone who has had vertigo has had experience with the importance of sight in helping to orient you. In the dark without much to focus on, most people with MD will find their balance impaired - often markedly so. One test of balance in the dark is to close your eyes, cross your arms over your chest, and try to march in place. Depending on how afflicted you are, you may find you are unable to do this without falling over, or at least thinking you are going to. Or if you are steady enough to stay upright, you may find yourself marching in a circle. If your balance is relatively intact, you'll march in place. (This is known as the Fukada Stepping Test.) In general, I became aware that I was much more likely to stumble and bump into things while walking around in dim light.

The strangest experience I've had, however, has only taken place outside in pitch darkness, and is different from staggering off balance or struggling to walk. While going from the house to the car, for instance, I have found myself walking sideways, feeling as if a magnet was pulling me strongly to one side, or as if I've been snagged by a fishing line and was being reeled in. I *had* to walk to the side – I couldn't stop it. The only way was to stand still – as long as I kept moving it was going to be sideways. I couldn't will it otherwise. It was not sickening or even alarming, not accompanied by the fear of falling. It did not even *feel* like a loss of balance – just a highly peculiar sense of being pulled by a mysterious power. Several other people reported having this same experience. In my case, this never happens except at night. Obviously, most of the foregoing are not experiences one might have during a severe vertigo attack, the kind that renders you unable to stand or walk unaided. They occur during the times when you are afflicted to a lesser extent, able to get around but not feeling fully well.

It is commonly understood that people will be slightly off balance when dizzy, and that drinking can also have a major effect on balance. Inebriated people sometimes walk with widely spaced feet to help with balance, or may be so affected that they cannot stand without assistance. Because of these well-known phenomena, it makes sense to everybody that an intense attack of vertigo will cause a person to be unable to stand or walk, or impair the ability to do so. However, interesting and subtler things happened to me during times when vertigo was milder – perhaps not even truly vertigo. I would be capable of going out for a walk, but it would be no pleasure, because I had to work at keeping myself walking straight the whole time. I could feel the motion of walking inside my head in a very active way, as if my brains were jouncing a bit with each step. Thus, I could keep my balance only with a conscious effort, and held my head as still as possible to reduce the jouncing. Probably because the motion seemed to churn up the vestibular system, walking for more than a few minutes would also tend to make me vaguely nauseated, a sensation that never left quickly. It was fatiguing for obvious reasons, and when I returned home, I would be sorry I went out. Again, others reported similar experiences.

When making my way casually around the house during such times, I might start to lurch to the side, and need to step out to keep from falling. I often bumped into walls. Any time I wasn't paying good attention to what I was doing, I was likely to stagger. As long as I was vigilant, such things could be pretty well controlled, but it's hard to keep a continual focus on something that usually operates automatically. And there were times when vigilance wasn't enough. One such time would be rounding a corner, and hence changing direction. My brains, which felt as if they had a Jello-like consistency, seemed to keep on going in the direction I'd first been headed. This would cause me to take a step or two more in the wrong direction, before I could get around the corner. The only way I could prevent this was to stop, think, and then switch direction from a motionless stance – a bit like the changing of the guard at the Tomb of the Unknown Soldier, certainly appearing a little odd to any onlooker.

Stairs are daunting to anyone feeling the least bit off balance. The care needed in going down seems pretty obvious – you hold the railing and watch your step. The difficulty in *climbing* stairs came as a surprise to me. You have to remember to keep yourself leaning forward, something people normally do without thinking. Otherwise, you run a risk of stepping backward into empty air to catch your balance, which would be potentially disastrous.

As mentioned, it seems as if vertigo is what causes disturbances in balance, since these disturbances always accompany it. Yet, for a person with vestibular problems, it is possible to have episodes of significantly poor balance when your head feels pretty ordinary, your muscles are as strong as they ever were, and proprioceptive sensors are presumably operating well. Sometimes such events are similar to those that occur because of vertigo, characterized by trying hard to maintain balance as you walk about, or suddenly losing balance and grasping something quickly to stay upright. Being clear-headed and stepping out of line feels stranger and harder to understand than losing balance while woozy. Since many people feel tired, or sick, or woozy, or nauseated for days after an acute attack, it may be that a continuation of balance problems hangs on in much the same way, and resolves gradually. It also seems that attacks of vertigo may be destructive to your inner ear over time, just as they can be to your hearing. It does seem to me that I have some permanent damage to my vestibular apparatus. But it is a subtle impairment, nothing that impacts significantly on the quality of my life. It means that unevenness in terrain may make me stumble more easily, or that I occasionally lurch to the side for no obvious reason. As long as I'm feeling well and focus on what I'm doing, it's not a problem. I can still engage in sports that challenge balance, such as skiing and surfing, and do about as well as I ever did. However, several people who had had intermittent attacks of vertigo over many years described to me ongoing and troubling problems with balance.

In considering whose balance was more likely to be affected *between* episodes of vertigo, diagnosis was not always a factor among the people I interviewed. Frieda, who is diagnosed with BPPV, had had rather brief episodes of vertigo over the course of many years. What was worse for her was the effect on her balance, with significant problems lingering for well over a week. She was unable to walk straight, would "bounce off the wall," felt out of control of her body which was "going where it wanted." (She also felt pulled to one side, as I described above.) She felt as if she were walking through a viscous substance, and was unable to walk at all on uneven ground. She also felt unable to drive, so that it was only possible for her to continue working because her office was next door and her job was sedentary. Thus, she was significantly impaired by something going awry in her vestibular system, but vertigo itself was the least of it.

Todd, who had been given a variety of diagnoses over about six years, also continued to experience problems with balance in the wake of a vertigo attack. He struggles to walk straight, and also feels pulled to one side. Within his head, he gets "roller coaster" feelings, and the sensation that his brains are sloshing. Cassie, whose present diagnosis is MD, hugs the wall as she walks for support, and needs to use a cane. She had to "relearn" walking and carrying. Roughly half of the people I interviewed gave similar reports, and it is clear that for some, the effort of maintaining balance can present a very major challenge, quite apart from actual vertigo. It is not always possible to make a neat delineation between the two, since they are usually intertwined, and because these experiences are deeply confusing and tend to blur together in people's minds.

To the extent that you can separate balance problems from vertigo, or in the case where a person's vestibular system has been damaged in such a way as to create ongoing balance issues, physical therapy may be helpful. Part of this may involve working to build up strength in feet and legs. But since actions performed by muscle systems also involve the proprioceptive system, the two cannot really be addressed separately. Physical therapists trained in vestibular rehabilitation will do more than just build up strength. They will guide you in becoming more conscious of your own body by increased awareness of how you use your eyes, move your head, your posture, the placement of feet and the way you move. You can modify some habits for the better, giving you better control. As noted earlier, activities such as dance, yoga, and tai chi may also be beneficial in this regard.

10

Ups and Downs with Vertigo – My Illness

Course of My Illness After the First Summer:
Mid-2006 to Early 2008

Remember that, in my life, vertigo began in February of 2006. This is a continuation of my saga.

After mid-July and for the remainder of 2006, I felt fairly well most of the time, and was able to lead my life pretty normally. But nearly every day, I did notice something associated with vertigo. Flashes of wooziness, stumbling off balance, pressure in my head or face, clicking sounds, zinging in my brains. Tipping my head back or lying down in bed could precipitate a brief sense of vertigo. When I returned to my job at the school in September, I found it more tiring than I ever had. Overall, however, the symptoms associated with vertigo were minor events, and were only an issue because they let me know my inner ear was still unstable. It seemed as if something was patiently waiting to catch me off guard and lay me low again. The whole autumn continued on in much the same way, with no disturbing bouts.

The year 2007 started out really well, with most of those daily reminders gone, such that I'd had seven months with no attacks, and I became hopeful that it might all be coming to an end. However, just over a year after my first attack, on a Sunday morning in March, and two days after a colonoscopy, I woke up with a new and startling variation on vertigo. When I opened my eyes to greet the day, they seemed to be flying all over the place, like a googly-eyed doll. I thought, "Wait a sec, just close your eyes for a moment and this will go away." But it didn't. This was different from before – the main thing I was aware of was the wild motion of my eyes. It wasn't that my brains were churning about or that the room was whirling around. Unlike in the past, as long as I kept my eyes shut, I felt pretty okay. I lay back down in my bed and went to sleep. When my husband came upstairs to investigate, he helped me get up to use the bathroom. That was trouble! My balance was terrible – I was unable to stand without assistance - and he had to hold me up while I shuffled along. And as soon as I started moving, I began vomiting. So I spent the next 36 hours in bed, luckily able to sleep most of the time, vomiting only when I moved.

The next day, about 36 hours later, I went back to the ENT again, and was again given a course of steroids. As had been true a year ago, I was uncertain whether to credit them with the gradual diminution of symptoms over the course of the week. And distressingly, the incident ushered in another several months of feeling unwell a great deal of the time. Now vertigo attacks were apt to begin when I woke up, and presented differently from the way they had previously. When I opened my eyes the objects in the room would slide sideways, right to left, repeatedly. In some ways, it was fascinating to observe how everything seemed to stay in the one place, yet move away – a real contradiction in perception. The sliding furniture phenomenon would diminish over an hour or two, but for the rest of the day I would feel woozy and nauseated and off balance, just as I had before. Even if there had been no vertigo in the morning, I was queasy and off balance a lot of the time. Sometimes I threw up. I was also fatigued most of the time and was often exhausted after a day at work. Food was unappealing and again I lost about 10 pounds. As before, I was worried about whether I could keep up with my job if this continued.

Near the end of spring, I wondered about getting another medical opinion. I called the health plan we belonged to and, in a hum-de-dum voice, a receptionist offered me an appointment for the end of the summer - no rush. I felt desperate when she said that – this had begun to seem like an emergency to me. The end of the summer was weeks and weeks away. I was barely getting through my life - there was must be *something* else to be done about it. Furthermore, what was happening to me didn't match the descriptions of MD I had read. According to those, I should have been having acute vertigo for a few hours, then waking up the following day feeling fine. But it was a rare day when I felt completely fine. I wasn't having disabling, on the ground, whirling vertigo – most days I could function – but my whole life was dominated by feeling awful most of the time, and the dread of feeling worse. And some days, I *did* have strong vertigo, coming and going at whim. It wasn't that I *wanted* to have a textbook case of MD, exactly. In fact, I hated the thought. But there was something weirdly disturbing about whatever-I-had being different from everyone else's. I couldn't imagine going on like this until the end of time.

We heard of a doctor who had recently come to our area who was said to specialize in inner ear disorders, and I was relieved to be offered an appointment with him in June. As it turned out, he was an otologist, a sub-specialty within the field of otolaryngology that focuses on the ear. ("ENTs" are otolaryngologists, who deal with noses and throats in addition to ears.) He ran a few tests, some that I'd had the year before. At our second meeting, he asked if I'd had another attack, and I didn't know how to answer him, since it seemed to me I was having some sort of attack almost all the time, even though it wasn't intense vertigo. I asked him about surgery, hoping it might be a cure, but he said he didn't do it unless the patient begged for it. I felt about ready to beg, but also thought his answer indicated he wasn't very enthusiastic about surgery. He recommended injections of steroid directly into the ear, which was a new idea to me, and seemed like a good thing to try – much less drastic than surgery. These were **transtympanic** injections, meaning that they went through the eardrum, so the steroids could be placed right by the inner ear. It sounds alarming, but was not really painful. The injections were done in July about two weeks apart. And maybe – or even probably – they did help. The rest of the summer really *was* better than the spring had been – I was way less tired, partly thanks to being on vacation – and the days were less permeated with symptoms. But I was likely to lose balance during rigorous exercise, and was upset that I had trouble on a beautiful long hike we took in the high mountains, needing to make sure I kept stumbling and lurching *toward* the rock face rising to my right so as to avoid toppling over the drop-off to my left. But that was the worst of it, and most other days went pretty well. I remained uncertain how much credit to give the injections, since the improvement was gradual rather than dramatic. Maybe the fickle pendulum was just swinging back of its own accord.

Whether it was due to the injections or the natural course of the condition, by about mid-September, my overall sense of wellbeing was much improved. I had much longer periods than I'd had earlier that year of feeling relatively normal. However, this is not to say that trouble with vertigo was behind me. Throughout the rest of the autumn, I had intermittent 3-4 day episodes when I felt woozy, nauseated, stumbled around (especially in the dark.) Generally, I found taking walks for exercise aggravated these feelings, as mentioned in the chapter on balance. In October, my sister arrived for a visit, and although I'd felt somewhat off the day before, we decided to hike up a local mountain. Some of the hike involved very uneven terrain, with boulders to clamber up and over, and some steep pitches. This turned out dreadfully! My head began to spin with the effort, I felt as if I might throw up, I could feel my eyes dance around, and had to sit down several times to avoid falling. Stopping like that did quiet my head after a few moments. The rest of the day was compromised because of this – I needed a nap and ate only toast for dinner.

Moving on to the next year, 2008, my story was similar to the previous year. During the first half of the year, I had several days every month wherein I would wake up to see the room sliding past me, or would have jumpy eyes. During those days, the perception of moving furniture would fade, usually within an hour, but then I would feel under the weather all day: woozy, somewhat nauseated, fatigued, dispirited. My balance would be noticeably poor. I didn't miss much work on account of it, but my overall level of energy and enthusiasm were diminished. I felt vigilant about the possibility of a disabling attack, and often down in the dumps. One big disappointment was missing a birthday party for one of my sisters in early July. The night before it, I was awakened by roller coaster sensations, swooping down and then up, over and over again on the left side of my head, which led to repeated vomiting. I was exhausted, nauseated, and way off balance the following day, in no condition to drive 100 miles to a happy celebration.

After one more episode in mid-July, most of the rest of 2008 was much improved. I still had woozy days, or pesky reminders of the condition sprinkled through various days, or staggered around in the dark, but the times when I felt impaired were only occasional. Although my tendency to become overtired continued, I began to think maybe the nightmare was fading away, just as I had thought the previous year.

In closing, I should note that I continued to take the diuretic hydrochlorothiazide daily over all this time. I pretty much stopped using either Meclizine or Valium, because they didn't seem to do that much good, and they tended to make me sleepy and tired. I felt tired anyway and didn't welcome more of it.

11

The Search for a Solution, Part I

To a person recently experiencing vertigo for the first time, worried about the impact it may have on his or her life, it is not reassuring to read about long struggles to find treatment, nor to see that for some the effects of the condition over time have been significant. However, in the course of my interviews the majority eventually experienced a reduction in symptoms or found ways to cope with them, such that they were gone or diminished, or so that life was normal most of the time. Some people did find a treatment that seemed to cure it. At the outset, it is truly hard to guess how it will evolve, so for quite a while a person may not know what s/he is dealing with. It is important to bear in mind it is common for the first attacks to be the worst ones.

Most of those who are experiencing vertigo, particularly if they've never heard of it before, are deeply anxious about what is going on. If they go to a doctor, they hope for a clear explanation and some treatment that makes the vertigo go away – quickly, and preferably forever. The Epley Maneuver, discussed in Chapter Seven, often *is* effective in alleviating vertigo immediately. But it is not always offered even if there is reason to suspect the patient has BPPV, which compounds the lonely feeling of being in the grip of the unknown. Hal, who profited from the Epley quite soon (about a month) after developing vertigo, commented that it did not seem to be part of routine ER procedures, and wondered why his primary doctor didn't seem to know anything about it. And of course if you have MD or vestibular neuritis, the Epley is not going to make a difference.

Apart from the anxiety and frustrations inherent in having a relatively mysterious condition, this chapter and the following will contain the stories of many who ultimately found a therapy that worked, or at least helped a lot, so that these people do not feel as if their lives are dominated or gravely altered by their condition. Sometimes these therapies ended up being in the mainstream, sometimes they were unconventional.

Struggles and Confusion with BPPV

The Epley Maneuver is a wonderful thing, often a definitive treatment for those who have uncomplicated BPPV, so one might expect that anyone given that diagnosis would be all set. Certainly the existence of the Epley must be the envy of nearly all who are diagnosed with MD. But as noted previously the maneuver is not always offered, possibly because there are still medical personnel who are unfamiliar with it or who have some reservations about it. If the diagnosis is unclear, doctors may hold back. Thus, many people told me stories of months or years of struggling with vertigo before encountering the Epley. On the other hand, I spoke with a few who experienced brief episodes of vertigo, who found those episodes so short or so occasional they didn't see much point in bothering about treatment. Some people resist trying the maneuver because they find it hard to face feeling worse during the procedure. But for most, a long delay before trying the Epley or a related maneuver means a long period of suffering, anxiety, and confusion.

The following two stories describe people who did not experience an enormously long period of distress after the onset of vertigo, but the roadblocks they encountered are instructive. In the first case, that of Helen, the Epley Maneuver was offered relatively soon, but the ensuing course of events was not entirely smooth. In the second case, Sylvia did not mind the fact that years elapsed before learning of the Epley, since her vertigo was not overwhelming.

Four years before we spoke, Helen, then 68 years old, woke up and saw the ceiling above her head moving. She closed her eyes but began to feel nauseated. She tried to get out of bed but couldn't stand, so crawled to the bathroom. After vomiting, she crawled back and lay in bed for about an hour, waiting to feel better. She finally woke her husband and he called an ambulance. They were both frightened, worried about a stroke, thinking that at their age anything could go wrong. Apart from ominous thoughts, the feeling itself was "unfamiliar and scary." At the ER, she was given a shot of Meclizine, and gradually the vomiting stopped and she began to feel better. The physician's assistant told her he was pretty sure she had BPPV, and performed the Epley Maneuver. She went home feeling better after three hours, but felt "dizzy and spaced out" for another two days, then was well for nearly 1½ years.

The next episode occurred during a yoga class, when she became suddenly overcome with vertigo and vomited on the floor. She again went to an ER, but this time the resident physician on duty refused to do the Epley and referred her to a neurologist. After an exam, where she learned her affected ear was on the right side, the Epley was done, but recuperation took much longer than before. She wondered if the delay in performing the maneuver was responsible for the delayed recovery time. For days, she felt shaky, had a "woozy head," kept wondering how much of this she might be imagining. Again, the symptoms cleared after a protracted period of time. After this, she and her husband learned how to do the Epley at home, and she says she brings the instructions, including a video of it, everywhere she goes.

During the first three years after Helen developed symptoms, she had three acute attacks and one mild one, and otherwise felt well. The fourth year was uneventful. During these years, she investigated other things that might help. Her internist referred her to a physical therapist in a vestibular clinic, where she learned some "habituation exercises," designed to help the eyes, brain, and ears work in concert. Later on, she also learned of a local ENT who specialized in vestibular disorders. She didn't encourage Helen to do the Epley on her own. Helen pointed out she had already been able to manage it with some success, and wanted the option of continuing to do it independently. The doctor relented and referred her to an audiologist who gave her training. Helen found her contact with the ENT very reassuring, especially since she was told the attacks would diminish with time.

Helen was clearly active in trying to figure out what was wrong and what she could do about it. She had been very distressed about her second visit to an ER, upon being told that they were not going to give her the Epley right away. Perhaps there are some conditions where it is contraindicated, but there is a good possibility that the resident was not really familiar with it. It is noteworthy that Helen's response to the Epley Maneuver was not the expected one. It took two days for it to take effect, rather than right away.

25 years ago, Sylvia, then in her mid-40s, was alone in the house. She woke up in the morning to a spinning world, was unable to walk, and crawled to the bathroom. A friend drove her to see an ENT, who told her that vertigo was not usually serious and mentioned BPPV to her as a possibility. To be on the safe side he ordered an MRI, which was normal. She was told she had "a benign virus in the brainstem." For two weeks, she was unable to go to work or to drive. She remembers holding onto walls as she made her way around the house. Despite regaining enough balance and clarity to go back to work, she did not feel herself for about six months.

Sylvia then had 15 years with no symptoms. When the vertigo returned, it was not as intense as it had been before. It was another seven years before the Epley was tried, to good effect. The PA (physician's assistant) who did it sent her home with papers from the Mayo Clinic instructing her on how to do it at home. Her husband helped her through it, and like Helen, she has found this to be very satisfactory. Unlike many others, Sylvia does not feel badly about how much time elapsed between the onset of vertigo and learning of the Epley

Maneuver. This is partly because her attacks have not been frequent, and partly because she felt so reassured upon finding out it was not a fatal condition. She wishes she had heard of BPPV before it began, so it would not have been so frightening, and likened it to the value of knowing ahead of time that flashing lights and "floaters" in the eyes are likely to occur as you age.

In the spring of 2012, Sylvia had an episode of vertigo that did not respond to the Epley Maneuver. She consulted with her doctor, who referred her to a physical therapist, not an option that had been previously mentioned to her. The PT determined that the problem was in the horizontal canal, rather than the anterior or posterior canals, where the problem typically occurs. Thus, she administered the "barbeque roll" (or Lempert Maneuver) mentioned in chapter seven. It did the trick, and was also something Sylvia was able to learn to do on her own at home.

The next two stories concern people who lived for a very long time with symptoms associated with vertigo without receiving any meaningful treatment or diagnosis. In both cases, their symptoms do not follow the classic descriptions of BPPV, and yet the Epley Maneuver has made all the difference in their lives.

About 40 years ago Frieda, now in her sixties, (mentioned in the chapter on balance) first found the world spinning when she moved her head to the left. Lying in bed was awful – if she rolled to the left side, vertigo woke her up. Lying down on her back made her feel sick. Or she would be sitting in a chair, and would suddenly need to grab hold of a nearby table, feeling as if she was falling off, when she was not. This happened occasionally throughout the day for about a week. She was able to walk, especially if she moved carefully. She thought something must be wrong with her ears, and went to see an ENT. He suggested Antivert, but also cautioned her that the medication might be addictive. He prescribed allergy pills and told her to move around as much as she could. Neither this appointment nor the advice felt helpful to her, but the symptoms began to subside on their own.

Years later, maybe 20 years ago, she began having such episodes more frequently, and with a much more profound impact on her daily life. The day before an attack she'd start feeling lethargic, weak, and nauseated. She couldn't walk straight – moving around "like a bumper car," sometimes bouncing off the walls. She had the experience of feeling drawn to one side by an unseen force. She was virtually unable to walk on uneven surfaces, and couldn't drive. Vertigo itself would be added to the experience the next day, but typically these were brief episodes. She would continue to feel lousy for over a week, as the symptoms gradually receded.

She went to a hearing and balance facility, where some "horrible tests" were administered. She has no recollection of receiving a diagnosis or treatment recommendations. She also spoke with her primary care doctor, who had little to offer. At some point allergy medication was again suggested. She didn't contact a doctor about it for a long time after that, seeing no point.

Then, in 2010, two years before I spoke with her, she had an intense episode that lasted for six weeks, and decided to see an ENT again. Upon examination, she told Frieda that she had nystagmus, which was something she'd never heard of. She felt "validated." The doctor seemed to believe something real was going on, that something concrete and identifiable was happening. The Epley Maneuver was performed, accompanied by the instruction to sleep sitting up for a couple if days. The vertigo left, but otherwise, she felt worse. The sickening feelings were more pronounced, and her balance was terrible, so much so that she wanted her previous symptoms back. This went on for another month and then eventually faded away. Her balance returned to normal, the attacks of vertigo disappeared, and over the subsequent three years, there has been no further trouble. It is unusual to experience heightened symptoms for weeks after the Epley, but this worked out well in the end and is thus a heartening story.

Frieda does not recall being given a diagnosis by any of the doctors she consulted. It was her son, a registered nurse, who suggested she might have BPPV. The history of her condition does not seem typical, as we have noted seems true for many afflicted with vertigo-related symptoms. The pervasiveness and severity of the effect on her balance, lasting days or weeks, is not mentioned often in the literature in connection with BPPV. On the other hand, her relatively brief moments of vertigo, set off by moving in a certain way, is very consistent with that diagnosis. Another symptom, not usually directly connected with BPPV, is that her hearing diminishes during an attack, but improves when it's over. This would usually be characteristic of MD. While there might have been understandable uncertainty about her diagnosis, for years she was offered no meaningful discussion, nor any seeming interest concerning what might account for her symptoms. The fact that she felt better after learning she had nystagmus is telling. Just knowing there was a name to describe some of what she was going through was an affirmation – it made her symptoms seem realer and more significant.

Yvonne's story is a particularly complicated one. Her symptoms don't sound like pure BPPV, and it may seem somewhat surprising that the Epley has worked well for her. In 1974, in her early twenties, and soon after the birth of her last child, she felt "the earth was opening up to swallow" her. She had no balance – couldn't stand or walk, "nothing was stable." If she kept her eyes and head very still, the spinning and/ or the dizziness would stop, but she felt terrible, very tired. She had no nausea, but did have diarrhea. She was afraid of lying down, because that made it all worse. She found it "big time scary," and felt "extremely abandoned." Her family thought she was going crazy and took her to a psychiatrist, who prescribed medication that she ended up throwing away.

These attacks would come and go, and later on another doctor tested her for a thyroid dysfunction but found nothing. In between she would feel normal, busy with her children. She lived in Puerto Rico, and moved to the States ten years later, in 1984. Here, the symptoms were attributed to anxiety, and she was treated for panic disorder with psychotherapy. She began experiencing brief episodes where she would "see the walls coming at [her]," or get a burst of dizziness when she moved her head a certain way. She felt doctors seemed annoyed by her, but met a primary care nurse who took a careful history, told her she had vertigo, and prescribed Meclizine. The medicine made her sleepy, but nonetheless she felt much better after taking it, as well as for having a professional take an interest.

In 2007, Yvonne was standing by the kitchen sink, when she felt as if she was going to fall, and went to bed. For two full weeks her balance was gone. Her daughter, who does medical research, suggested going to the Massachusetts Eye and Ear Infirmary, a drive of nearly two hours from home. There, an MRI was done, and the doctor told her about crystals moving in her inner ear. This was a huge relief to her, and she no longer thought she was "going nuts." He performed the Epley Maneuver, and her symptoms cleared.

She went home feeling light hearted - finally, she had an explanation that made sense and a treatment that actually helped. But unfortunately, that did not turn out to be the end of her story. Late in the year 2010, now in her late 50s, she began having sudden episodes she described as "going down to the floor," wherein she would have to grab onto something stable to keep from keeling over, in addition to her previous symptoms. During such episodes, her blood pressure would rise and her heart would pound. Over the subsequent two years, she has experienced pressure, and a "ringing, flapping" sound in her ears as well. She's been tested several times and there is no hearing loss. When Yvonne gets severe attacks, her husband takes her back to Boston, a drive of two hours, for the Epley Maneuver. They had not been successful in learning to do it at home, and could not find any facility near where they live that offered it.

Yvonne has raised her family, studied for an advanced degree, and has a rewarding job. This condition has been an intermittent disruption to her life, but has not defined it. She feels enormously grateful that she now has some understanding of what's going wrong, and has found treatment for it even if it is not a cure. Nonetheless, she had to wait about 30 years before that happened, years punctuated by overwhelming symptoms and fears about what could be wrong with her. The fact that no professional thought in terms of a vestibular problem for at least ten years, and that it was over 30 years until any meaningful treatment for it was offered, is testament to the struggle so many have had in finding understanding from those in the medical professions.

Most people are frightened when they first experience vertigo, especially if it lasts for a few hours and is severe and disabling. People who have brief spurts of vertigo may feel some alarm, but generally are less concerned and may not pursue much treatment over the years. People who respond to the Epley Maneuver soon after vertigo begins tend to feel – and be – much less afflicted by their condition. A few people I interviewed understood they had an inner ear disorder but had never been offered the Epley. They had been given Antivert or Meclizine and were content to sleep off the vertigo, appearing to view it as a nuisance similar to the flu – a condition that knocks you out of commission for a while, then goes away. In general, people with BPPV who have responded positively to the Epley are not likely to be on a continuing quest for a cure.

According to the dizziness-and-balance website,[57] 20% of patients do not find the Epley Maneuver to be effective. For them, a surgical procedure called "**posterior canal plugging**" may be an option. Such surgery would be offered only after the Epley has been tried several times and the patient has suffered at least a year with vertigo. The surgery carries with it a 3% risk of damage to hearing, and is not something to be undertaken lightly. A person might need to travel to a distant city in order to have it, since it's described as a specialty not available everywhere. I didn't talk with anyone who had done this, and so have no first hand account.

Struggles and Confusion with Menière's Disease

MD is apt to cause more trouble than BPPV does, in some measure due to the fact that there is no treatment similar to the Epley Maneuver that can put a stop to an attack. Sometimes it does fade away and disappear on its own, but usually it is a remitting condition.

---

57   www.dizziness-and-balance.com Affiliated with the Chicago Dizziness & Hearing Center.

Sometimes long periods of time elapse in between episodes, so that it doesn't loom very large in the person's mind, except as a cloud lingering in the background. The longer a person goes without an attack, the less likely he or she is to be deeply concerned about it. Sometimes people see a significant improvement by following the dietary restrictions, possibly combined with other life style changes, or the use of some medication. These people, and those whose attacks are relatively mild and infrequent, can usually come to accept with some measure of peace the likelihood of vertigo being an occasional event as their lives unfold. This section is mostly about people for whom the initial treatments, described in Chapter Six, did not make enough difference, and who experienced an ongoing struggle with MD until finding something that did help.

Surgical Interventions for Menière's Disease

In reading up on the secondary or alternative treatments for MD, in most sources you will see listed the possibility of gentamicin injections or of surgery. Both surgery and gentamicin pose significant risks. Apart from the risk inherent in surgery itself, vestibular nerve section, labyrinthectomy, and gentamicin injections are all thought of as "destructive," since the intent is to cure vertigo by destroying the balance function of the inner ear. The "good" ear will then take over that function, after a period of adjustment that is likely to take some time. With regard to surgery, the types most often mentioned are:

- Endolymphatic sac decompression
- Endolymphatic shunt
- Vestibular nerve section
- Labyrinthectomy

Endolymphatic sac decompression and the placement of an endolymphatic shunt are sometimes done together. For the decompression, a small particle of bone is removed to relieve pressure on the endolympatic sac. For the shunt, a tiny tube is put in place to operate as a permanent drain to prevent the endolymph from building up and creating "hydrops." or excess endolymph. Both sound as if they make sense, since MD is believed to be the result of pressure within the inner ear that is caused by high levels of endolymph. This is, of course, surgery near the brain, and would not be advisable without very careful consideration. It would seem more clearly worth it if there were strong evidence that it was effective a significant percentage of the time. It does work for some, but for many, symptoms recur after a few years. Although there have been studies that variously show a 50-75 percent success rate, it does not appear to be embraced with enthusiasm by the ENT community. According to an article published in 2010 by Steven Telian[58] it is recommended only for intractable vertigo. It could be that the use of low dosage gentamicin (discussed later on in this chapter) may now be seen as a less drastic alternative. I only spoke to two people who had had the surgery done.

One was Jake, who was hit strongly with vertigo at age 47, in 2009. The onset was gradual, but after a day or so, everything in his vision was moving, sliding, on a roller coaster. He was unable to stand. This was accompanied by headaches and a buzzing tinnitus that tended to begin near the end of the day and last for around six hours. He'd wake up in the morning somewhat better, only to experience another recurrence in the afternoon. For two months, he never felt well. When he was able to walk, he sustained bruises from bumping into things. He noticed diminished hearing and experienced some double vision as well during the first two weeks. He met with a doctor when the symptoms persisted, who tried the Epley – to no avail. The doctor prescribed a course of steroids, which made no difference. Because of the concern about hearing, Jake requested an appointment with an ENT, who confirmed the loss, and ordered an MRI to rule out other possibilities. A diagnosis of MD was made, and the standard dietary recommendations for treating it, along with a diuretic, were recommended. In

---

58   Telian, Steven. "Surgery for Vestibular Disorders." Cummings Otolaryngology: Head & Neck Surgery. 5th Ed., 2010. Paul Flint, Editor.

addition, balance therapy was suggested, which did help some, as did Meclizine. He felt somewhat better, but not wholly free of symptoms, and had ongoing trouble concentrating. "I couldn't keep a thought in my head." He had to leave his job as a retail pharmacist, since he was unable to stand up all day.

Jake's MRI had shown that he had a deviated septum, so his doctor suggested that, during surgery to correct that, an endolymphatic shunt also be performed. This double surgery was performed in early 2010 about eight months after the onset of vertigo. Joe considers the results to be good but not perfect. Most of his dizziness resolved, his brain "works better," and the pressure in his head eased. He still has some intermittent balance problems. As it has turned out, he likes his new job better than the previous one. And he can do most of the things he enjoys, but "chooses not to water ski." Interestingly, about a year and a half after the surgery he developed a severe case of poison ivy, and was put on steroids. During that time, *all* symptoms left! This, despite the fact that earlier on, steroids had not helped. Now he wished he could stay on steroids forever, but of course that was not to be. Jake continues to observe a low salt diet, take hydrochlorothiazide, and expects he will always have some low level symptoms. As he says, "They're tolerable."

On the other hand, Jody (mentioned in Chapter 3) had endolymphatic shunt surgery that "didn't change a thing." She'd had some vestibular symptoms most of her life, especially tinnitus, and had always been clumsy, bumping into things and tripping easily. Then, at age 33, in 1994, she had an intense vertigo attack that lasted for days. She crawled back and forth between her bed and the bathroom, vomiting frequently. The room spun, mostly left to right, and her eyes "bounced around." Both her ears hurt. As soon as she was able to get up, a friend took her to a doctor, who thought she had an ear infection and prescribed antibiotics.

Although she began to feel a bit better, she returned to the doctor within a few weeks. She still had pain in her ears, the tinnitus was louder, she had several "mini-whirls" a day, and was exhausted. The doctor changed the antibiotic. She went on as before, until in 1996, when she suffered an attack that lasted most of the day. She returned to her doctor and insisted on a referral to a specialist. She lived relatively near a university hospital, where she received a thorough evaluation, including an MRI and a hearing test, which showed a small loss. The doctor there told her she had MD, and took a lot of time to explain it. While it made sense to her, she also felt angry about the diagnosis and found it hard to accept. The standard recommendations and prescriptions were provided. Despite all that she got worse. She would get 10 to 15 small attacks a day, and vertigo lasting several hours at night. In 2001, a steroid (Decadron) injection was tried, to no avail. Soon after that and seven years after this had all begun, an endolymphatic shunt was done in June of 2001. Because that was unsuccessful, a vestibular nerve section, meaning that the nerve itself would be severed, was suggested a year later. She "freaked out" at the thought of this, and refused it. About a year later, she had to stop working and go on disability. Jody's story is a tough one, and we will return to it later, during the discussion of more unusual therapies.

Neither of the two stories above can stand as proof concerning the advisability, or lack thereof, for undergoing these surgeries. You need to look for large population studies for that, although they don't seem to provide a strong case for a dependably favorable outcome.

One famous person for whom it *was* successful is Alan Shepard, the astronaut, who developed vertigo in 1963. He had an endolympatic shunt put in place and was able to fly on the Apollo 14 mission to the moon in 1971. Shepard's attacks of "intense, debilitating dizziness" accompanied by "terrible ringing in his ears" usually hit in the morning.[59] They began around the time President Kennedy was developing support for the race to put a man on the moon – the Gemini flights. Shepard was among those chosen to participate in this, but so far he had told no one associated with NASA about the vertigo. A Christian Scientist, he had been trying to take care of it himself, although he did relent to the extent of visiting a doctor and receiving some medication

---

59   The information in these paragraphs is summarized from Light This Candle: The Life & Times of Alan Shepard by Neal Thompson.

and vitamins. Despite that, his episodes persisted, becoming more frequent and more dramatic. He felt well between the attacks, which enabled him to keep them secret, but was aware of the risk they posed to his dream of lifting off into space. Apparently he kept hoping the vertigo would just disappear.

Shepard was forced to admit he was in the grip of a serious condition after an attack that took place in public. While in the midst of giving a lecture he began to lose his balance, clung to the podium but could not continue, and had to be helped off the stage. Later on that day, he consulted with a NASA physician, and after a series of tests, was told he was grounded. His diagnosis was Menière's Disease. He lost his place on the Gemini flight but stayed on as an advisor and administrator for the program. For a while there was some improvement in his symptoms, but the condition returned strongly in 1968. By then he had lost most of the hearing in his left ear. That year he heard about Dr. William House in Los Angeles who had performed a few shunt surgeries. Shepard decided to go for it, knowing he was taking a risk and that results were not guaranteed. A few months after the operation he found himself to be free of symptoms, and was so confident that he requested being reinstated as a full-fledged astronaut. Before giving him the command of Apollo 14, NASA tried whirling him around and dunking him underwater to check for signs of underlying disease. He left earth for his trip to the moon on January 31, 1971, where he is known for driving a golf ball through the vacuum of space, sending it soaring over the pocked and ghostly landscape. I was unable to learn whether he ever had another episode of vertigo, and I doubt he would have publicized it if he had. But he, his fellow astronauts, and NASA were fortunate that he was one of the lucky ones for whom the surgery made all the difference.

Vestibular nerve section (VNS) is an operation wherein the vestibular nerve that serves the affected ear is cut, so that signals about balance are stopped from reaching the brain stem. The idea is that the unaffected ear will take over the balance functions. It is a very final act, and carries some risk of inadvertently cutting the auditory portion of the nerve as well. It is thought of as a last resort, and I spoke with no one who had undergone this procedure.

A labyrinthectomy consists of removing the entire inner ear on the affected side, again with the idea that the other side will assume not only the balance functioning but the hearing as well, since the operation renders the patient completely deaf on that side. It is not performed unless the person has lost all, or nearly all, hearing in that ear already. As with the VNS, I spoke with no one who had had this done.

Gentamicin Injections

The hoped for result of providing gentamicin injections is to destroy the balance function of the inner ear, while leaving the hearing function intact. The injections do not have the risks inherent in surgery such as the tricky work of cutting into the brain or the head, or even the anesthesia itself. But they do carry the risk of damaging or destroying hearing, although more recently developed procedures for administering it have diminished that risk. For people whose hearing is already compromised, and/or who are miserable with vertigo, the risk may not matter much.

Gentamicin is actually an antibiotic, which was discovered (quite by accident, as so many things are) to destroy the cilia in the inner ear. If *all* the cilia are destroyed, neither the cochlea nor the vestibular system will convey any more information to the brain. In order to treat vertigo, gentamicin is administered by an injection through the eardrum (transtympanically) into the inner ear. Years ago, when this was first done, the patient would lose all inner ear functioning, including hearing, as in a labyrinthectomy. The advantage was that the procedure itself, although labeled "destructive," was less drastic than surgery. But more recently, it has been found that using a series of much smaller doses carries less risk to the hearing, while still achieving the desired effect of deadening the balance receptors in the vestibule. Remember the idea is that after balance regulation in the affected ear is lost, the inner ear on the unaffected side will take over. As with surgery, the procedure is irreversible. You can't change your mind later. Gentamicin is very effective in stopping bouts of vertigo, although it is not 100% - not quite a miracle drug. Some patients may need to receive booster shots of it over time. Depending on who's doing the research, the results look a little bit different.

After receiving gentamicin treatment, or after surgical removal of the balance function, patients will experience a few days – sometimes longer – of vertigo until the brain adapts and the other ear takes over. Apart from the risk to hearing, one problem with any of the procedures that destroy the inner ear is the possibility that a person could develop vertigo in the second ear later on. This happens in about 10-20% of the people diagnosed with MD. Two people in my sample considered getting the injections, but decided against it because of the threat to hearing. Only one of the people with whom I spoke had gentamicin injections – sometimes colloquially referred to as "gent." No one moves on to any of these procedures, nor to anything unusual, without some sense of desperation.

Sally, who ultimately was treated with gentamicin, had a somewhat gradual experience with vertigo in the beginning, but it became so overwhelming after about a year that it was unclear whether or not she could keep her job. In October of 2008, at age 56, she'd had a hysterectomy, and the day afterward she noticed fullness in one ear, along with some loss of hearing, slight tinnitus, and mild dizziness. A few months later, in March of 2009, while outside in the cold wind, she had a bouncing feeling, her eyes began moving all around, the world seemed to be twirling. She struggled to get herself inside, and sat in a room with the lights off, waiting for it to go away. She was terrified and tried to talk herself down, telling herself she was okay, it would go away, it was a fluke. And it did go away – but only for two months, until May. Then she felt a "roar rushing into [her] head, experienced very strong vertigo, was unable to stand, and began fierce vomiting." That episode lasted three hours, after which her husband took her to see an ENT. He put her on a diuretic. She felt really terrible for about a month after that. Her blood pressure was very low, she felt as if she "had no air in [her] body," and was very frightened about what was wrong and what was going to happen to her.

Attacks occurred from time to time over the next year, becoming both more frequent and more intense. Despite this, she hung in with her job as a teacher, a job she loved. Sometime during that year, she saw someone who specialized in the inner ear, and had a thorough evaluation, including an MRI, hearing tests, caloric tests, and so forth. The diagnosis of MD was made, and a low salt, no caffeine, no alcohol diet recommended, along with Triamterene (a diuretic). Flavanoids were also suggested. Meclizine too, but it didn't help much. She met with the doctor every three months, but her symptoms kept getting worse. She lost more hearing, down about 50%, and the tinnitus grew louder, sounding like "aluminum foil rubbing on steel." During vertigo, everything was "bouncy, rubbery," and she couldn't stay balanced even sitting on a chair. By the fall of 2010 she was getting attacks every day, and the doctor said there was nothing more he could do. Despite much more regular care than most people reported to me, she never felt the doctor understood what was going on, and was deeply discouraged.

A friend told Sally about a clinic in Florida that offered treatment for MD, called the Silverstein Institute. She called there in October, and went down (from New England) for treatment in January of 2011. This institute offers gentamicin, administered in slow doses, via a "MicroWick" inserted through the eardrum. She went down there for more than two weeks, receiving four doses in all. By the time she arrived, MD had damaged her hearing such that she had an 85% loss, so she wasn't very concerned about the possibility that gentamicin might hurt it. She thought of it as already ruined, and anything that could take away the vertigo that dominated her life was worth it. The treatment gave her very strong vertigo, accompanied as always by compromised balance and vomiting. She went home after two weeks, but it took about a month to feel well. Since that time, a few months before we spoke, she's been feeling increasingly better, happy and amazed by the effectiveness of the treatment. She says she has her life back.

The MicroWick, which is a patented device, is a tiny tube that is left in place in the eardrum, to be removed when treatment is completed. It eliminates the necessity of puncturing the eardrum repeatedly if several applications of gentamicin are to be administered. However, according to Timothy Hain, MD, of the Hearing and Balance Center in Chicago, two small doses of gentamicin injected a month apart are usually sufficient. This is something that could presumably be done by most otologists. According to Dr. Hain, administering the medication in small doses makes the risk to hearing a very small one. To quote him, "Usually it is best to go somewhere else if your otologist suggests you will need four injections." People who live near a major medical center will probably be able to find a practitioner familiar with low dose gentamicin, and others may find a trip to such a medical center worthwhile. Anyone considering it would do well to read Dr. Hain's website on the subject,[60] as well as the website for the Silverstein Institute.[61]

Sally is the only person with whom I spoke who was treated with gentamicin. One other person reviewed it with his doctor, but they decided against it since his hearing was already compromised in both ears. No doctor wants to do anything that could damage hearing without extremely good reason. But even with the understandable need for caution, it is surprising that it hadn't been brought up for consideration in several other cases, including that of Sally's herself. (Remember that gentamicin was not suggested to Sally by a doctor, not even after he had told her there was nothing more he could do for her.) A small number of the people I interviewed continue to be impaired by MD, to the extent of having to cut back on work, change jobs, or go on disability. Some feel lousy at least as much of the time as they feel well. It is unclear why low dose gentamicin hasn't been presented to more patients as a backup therapy. Of course doctors want to prevent deafness, but being significantly affected by vertigo is also a serious condition that can be as handicapping in its own way as is defective hearing.

---

60   www.dizziness-and-balance/treatment/ttg.html (The problem with recommending a specific web site is that it may change by the time this reaches a reader. However, it is likely that the basic web site [dizziness-and-balance] will be available for quite a long time. Overall, this website is one of the clearest for information on vertigo related questions.)

61   Simply google Silverstein Institute.

Gentamicin is risky, but I've wondered if the word about the relative safety of low dose gentamicin hasn't gotten out yet. I should note here that many otolaryngologists might feel indignant at the assertion that physicians are averse to the use of gentamicin, if they are among those who do administer it. I based the statement largely on the experiences of my interviewees.

Within my sample of about 50 people, many - but not all – had received, at least eventually, the "front line" treatments for BPPV and MD that are most commonly believed to be therapeutic. Most people diagnosed with BPPV found the Epley Maneuver to be either helpful or curative, while the verdict on the usefulness of standard treatments for MD was more mixed.[62] Those who move on to consider surgery or gentamicin are those who continue to feel very impacted by their symptoms. But surgery and gentamicin do not seem to be enthusiastically embraced by many medical practitioners either. Particularly with regard to gentamicin, this reluctance may seem surprising, since widely accepted solutions to the problem of intransigent vertigo symptoms don't seem to exist. In the next chapter, we will look at some outlying treatments tried by some of the people I interviewed.

---

62   None of this has any statistical significance. Since my sample is so small, it could be very skewed. It's possible that the majority of those with MD do find the basic treatment regimen helpful.

12

Search for a Solution – Less Conventional Treatment

Looking into Less Conventional Treatments

There are many treatments people have tried, some with medical backing, some on their own, to try to ratchet down - or better yet stop - repeated episodes of vertigo. I won't be able to cover everything, but do want to discuss therapies that made a difference in the lives of those I interviewed. Hearing a success story first hand makes an impression, and I've needed to remind myself that you need way more than one success for a treatment to become accepted practice (more like a few hundred). And with any disease, any drug or therapeutic procedure, there are countless failures or successes that seem unexplained. In leafing through popular magazines or checking out Internet sites, you will find stories about miracle drugs, prayer, special diets, and so forth that have worked wonders for all sorts of ailments. A drug or procedure that works on one person may make no difference on others, even if the value of that drug is supported by scientific research. If common therapies haven't helped, people who feel desperate are apt to spend money and time on treatments that have little support in the medical community, or treatments that may seem, at first blush, to be based on little more than superstition. But sometimes these do help.

Anti-Viral Medication

In the medical world there is informed speculation, but not overall consensus, that MD may initially be caused by a virus. Some have also proposed that this may be true of BPPV as well. However, it is generally accepted that a virus may be the causative agent in labyrinthitis and VN, which are likely to strike only once. Richard Gacek, MD, an otolaryngologist at the University of Massachusetts Medical School, has written several articles on the possible viral origins of MD. One article in particular, published **in** 2009,[63] describes a study in which 121 patients with a diagnosis of either MD or vestibular neuritis (VN) were placed on a course of acyclovir, an antiviral medication. The response to the treatment was impressive, with vertigo "completely controlled in 73 of 86 patients with VN (86%) and in 32 of 35 patients with MD (91%)." Tinnitus improved in about half the group, but no reversal of hearing loss was seen. The virus, once having gained entrance, may be latent within the nerve fibers, restrained by the immune system, erupting from time to time in an attack of vertigo. Dr. Gacek's articles sounded persuasive to me, but apparently have not found wide acceptance in the medical community, which means that prescribing antiviral medication remains an unconventional treatment.

Lila, first diagnosed as having MD, was profoundly affected by her symptoms over a period of 14 years, until she was prescribed a course of Valtrex, a brand name for another type of anti-viral medication (Valacyclovir.[64]) She first developed symptoms of vertigo in 1997 at age 39, which initially felt like a panic attack to her. Her heart was pounding, it was hard to breathe, she couldn't walk without holding on, she kept vomiting. She described vertigo as the "insides of my head going around, Jello brains, every body motion magnified. Objects slid past." Her condition was perceived by her doctor as having a psychological cause, with stress being the culprit. A tranquillizer was prescribed, along with Meclizine. These made matters worse. After around two

---

63   Gacek, R.; "Meniere's Disease is a Viral Neuropathy," Journal for Otorhinolaryngology, ORL 2009;71:78-86, January 10, 2009

64   Valacyclovir and acyclovir are closely related medications. The former provides a stronger dose.

months of this she had to quit her job, and the doctor referred her to a psychiatrist, who gave her antidepressants, which were "no help at all." It took about 18 months for the vertigo to leave. She found a new job, and her health was normal for seven years, until 2004. "Then, bam! It happened again." This time she saw an ENT, who diagnosed her with MD, and gave dietary recommendations and a diuretic. But as before, the symptoms lasted for many months, months wherein her eyes flew around, objects slid past or bounced, the insides of her head spun. Sometimes she saw double. Again she had to leave her job. Again the symptoms slowly resolved and she felt well for another few years, although she noticed her balance had become somewhat impaired even when she felt fine. In 2006, she was able to return to work and again obtained a new job.

All was well until June 2011, when she walked into work, and without any prelude it hit her once more. She found some new doctors, and also received a new diagnosis, of bilateral[65] vestibular loss. Physical therapy was recommended but it didn't help much. She had to use a wheelchair to get around - "either that or crawl." She was bothered by tinnitus, something she hadn't noticed much before, which presented as a "bad hum" like an insect buzzing by her eardrum. She was unable to read because the print jumped. She was allowed to take a medical leave from her job, so this time she didn't lose her position.

In October of 2011, she went to see Dr. Gacek. He also found her impairment to be bilateral – in both ears – and diagnosed her as having vestibular neuritis, or VN, rather than MD. He placed her on a course of Valtrex. Within four weeks the vertigo was better, and by six it was gone. Her head felt normal, her energy and spirits much revived. However, her balance was still so poor she was unable to walk. It took another few weeks before her condition improved enough to allow her to return to work – after a nine month absence - although she needed the support of a cane if she had to walk any distance. If vertigo attacks have actually destroyed some nerve endings or cilia associated with hearing or balance, they cannot be restored. Nonetheless, she is

---

65   Bilateral means both sides.

pleased and is hopeful that more of her balance will return. She was told it could take a year.

Lila, although diagnosed with MD at one time, ended up with a diagnosis of VN. It is easy to see the overlap in symptomatology and the confusion that can ensue in arriving at a diagnosis. Two of her vision related experiences would usually be thought of as originating in the brain or nervous system, and would not be considered likely to occur in MD or BPPV. One is double vision (technical word: **diplopia**), which she developed in June 2011, at the time of her third major attack. The other is her experience of lying on the couch and noticing a picture on the wall breaking up into small pieces like a kaleidoscope, a phenomenon also more attributable to symptoms of central[66] origin. In thinking about a differential diagnosis, she stated she'd not suffered the hearing impairment that would have been expected for MD, although there was some loss for higher frequency sounds.

Brent's first attack of vertigo occurred in his early 40s, which was followed by 15 normal years. At age 58, vertigo returned. He would have a series of attacks for a few months, then a long respite. He would get a little warning before the vertigo became full-blown, but once it did, it would be difficult or impossible to stand. He would be lying on the floor with intense spinning – everything in the room moving around – and tended to vomit repeatedly for about half an hour. When the violent gastro-intestinal reaction subsided he could sleep for a couple of hours, and wake up tired and feeling weak but free of vertigo. The next day he'd be fine. The reader will note that this sounds like the "classic" description of MD.

Sometime prior to the return of vertigo attacks, he'd begun seeing an ENT because of a hearing loss, who told him the attacks "could be" due to MD. Life went along, with clusters of attacks followed by fallow periods, for about three years. In 2008, he had an attack that was atypical, in that it lasted two days. He was seen at a major teaching hospital, and had to be rolled into the clinic in a wheelchair. He reported that the doctor was delighted to see someone actually in the throes of an

---

66   Remember "central" means within the brain or nervous system.

episode, since usually by the time patients get to him they've largely recovered. At this visit, after some tests, he left with the diagnosis of MD.

For the next three years, Brent had some nice long periods free of vertigo, although it always returned from time to time, mostly following its pattern of an acute attack that resolved in a few hours. He came to think of it as an infirmity he could put up with, especially when he thought about the things many others in their sixties had to deal with. January of 2012 changed that, ushered in by a drop attack. He was walking in a park with his wife and was not hurt by the fall, but unable to stand afterward. She was able to get him to the car and bring him home, where he spent the rest of the day on the bathroom floor, intermittently throwing up and sleeping, waiting for it to go away. The next several months were sheer hell, with attacks happening as often as once a week, and sometimes twice in one day – in a restaurant, while visiting friends, any old time. Brent felt reluctant to leave the house, thought it was no longer safe to drive. His sense of himself as a person who coped well, who was self-reliant and confident, was shaken. He returned to the hospital in August to discuss the possibility of a trial on anti-viral medication. His doctor there said there had been no controlled studies of such medication, and didn't feel he could recommend it. He did discuss the possibility of an injection of gentamicin, but Brent was put off by his description of how horrible you would feel for several days following the treatment. He decided to contact Dr. Gacek on his own.

In September of 2012, Brent's wife drove him to Worcester to meet with Dr. Gacek. The doctor agreed to prescribe the antiviral medication Valtrex (valacyclovir,) and placed him on a regimen that called for a greater number of pills per day for the first three weeks, then gradually tapering down. At about three weeks, Brent experienced a small attack (no vomiting, no lying on the floor) that lasted less than an hour. Some adjustments were made to the dosing, and by the end of two months, he was taking only one pill a day, which he expected to do for the rest of his life. After seven months, he felt completely asymptomatic, and confident about the value of Valtrex, even though he also was aware of the enormous unpredictability of MD. Very unfortunately, about a month after feeling the disease was behind him, Brent developed another series of attacks, with vertigo arriving roughly once a week.[67] Some attacks were worse than others – the "dividing line" would be how long they lasted and whether they led to vomiting. His plan was to wait a few weeks to see how things played out before returning to consult with the doctor. As is so often lamentably true with vertigo, especially MD, it's impossible to know whether the course of Valtrex did any good, or whether it was coincident with an improvement that might have occurred anyway. The only thing that *is* clear is that it didn't turn out to be the magic bullet that eliminated the condition once and for all.

Dr. Gacek, with whom I spoke, has considerable conviction about the efficacy of antiviral medication. He believes it's advisable to pre-scribe it right up front, as soon as a diagnosis of MD or VN is made, in the hope of nipping the disease in the bud. He emphasized that there is no point in starting a course of the medication during a quiet period, indicating that it needs to be initiated during an attack when the virus is active. Once begun, a person might remain on the drug for a long period of time. It is unclear to me why little interest in using antivirals has been shown by the medical community, even if Gacek's studies are thought to be insufficiently rigorous, as some have said. You would think others would be interested in replicating his studies or designing their own.[68] Most of us who struggle with MD would be thrilled to take medication that would either eliminate or reduce the symptoms of MD. If it worked, it would be by far the most palatable alternative of any. Of course, from a historical perspective, it has often required decades for new therapies or approaches to illness to catch on – just as it does for new ideas about anything that don't fit into the current understanding.

Lack of proof of the presence of virus is another argument against

---

67 It's important to bear in mind that neither Brent's nor Lila's experience proves anything one way or the other about the value of Valtrex. Only large population studies can do that. It's my understanding that the medication is low risk.

68 At the time of this writing (2013), two studies were in progress, one at the Tehran University of Medical Science, and the other at the University of Sydney. There may well be more.

the idea. Relative to that, I have seen at least two conflicting studies, both investigations into whether signs of viral DNA can be found within the inner ear. In an article found on UpToDate, a website, it is said that no DNA for viruses have been found in endolymph recovered during surgery.[69] Yet another article stated that, in a study of ten MD patients who had endolymphatic shunt surgery, there *were* signs of antibodies for several viruses.[70] Thus, like so many other things concerning our vestibular diseases, there remains a mystery.

Craniosacral Manipulation

I return now to the story of Jody, discussed in Chapter 11, who had such a terrible time getting a diagnosis at first, and then did not find any treatment that helped, not even the endolymphatic shunt surgery she had. After she turned down further surgery (a vestibular nerve section), her doctor said he had nothing more to offer. She was able to find another doctor, one who was interested in MD, and who was willing to support her interest in trying unconventional approaches. Like a couple of other people I spoke with, she tried acupuncture for a while, twice a week, but found no obvious benefit. What did help her was **craniosacral manipulation**. With this, her tinnitus was gone, as was pain in her ears, and she had fewer vertigo attacks. It wasn't perfect, but it was the best she'd felt in years. Her insurance covered the treatment for two years, but then refused further coverage, even though her doctor contacted the company to confirm its value for Jody. Since then, she's lived with a chronic condition, unable to drive, can't tolerate noisy places, reacts with a "mini-whirl" to strong scents (such as the laundry aisle in the supermarket), worries about when she will be hit with the next attack.

Jody is one of two people I interviewed who tried craniosacral manipulation (CST,) and both received some benefit from it. In addition, I

69   Dinces, Elizabeth & Rauch, Steven; "Menière's Disease: Literature Review through 3/13" UpToDate

70   Yazawa Y, et al, "Detection of Viral DNA in the Endolymphatic Sac in Meniere's Disease by in situ Hybridization,"ORL, vol.65, no.3, 2003. Evidence for the presence of the Epstein-Barr, the cytomegalovirus, and the varicella-zoster (the chicken pox virus) was detected.

read a testimonial on the Internet by another person who said her vertigo attacks were actually cured by it. I also spoke with a friend who had been suffering, for over three months, from double vision caused by a palsy of the fourth cranial nerve.[71] Her symptoms seemed to have been cured by the method – or at least she recovered during the time she was having the treatments. All that said, it's not clear to me why craniosacral manipulation should make a difference. The therapist places his or her hands on the head to detect rhythms within. Gentle manipulation and pressure is given, with attention to each individual bone in the head, and their articulation with each other and with the spine, among other areas. It is intended to release tension and restore natural rhythms. The head may be tilted at different angles. It is a peaceful and soothing experience. The treatment method was developed by an osteopath, John Upledger, in 1970, who based it on earlier work done by William Sutherland (also an osteopath) around 1900, and is defined by Wikipedia as an "alternative medical therapy used by osteopaths, chiropractors, massage therapists, and naturopaths."[72] The article goes on to quote sources saying there has been "no valid scientific evidence that CST provides a benefit to patients." However, there's no arguing with success, as they say, and if something works for a given individual, especially one with a debilitating chronic condition, it seems wrong that insurance would cease to cover it.

Meniett Device

One possible treatment that you are likely to come across while searching through the Internet is the Meniett Device. It requires minor surgery to perform a **tympanostomy** (similar to the procedure for inserting the MicroWick), perforating the eardrum in order to put a tiny grommet in place. The Meniett is a small electronic device with a long flexible tube that leads to a special plug you place in your ear. The device emits controlled "micropulses" that pass through the grommet to exert gentle pressure on the round window leading to the inner ear. According to Dr. Ed Cheung, who offers a series of lectures on Youtube

---

71   The 4th cranial nerve is the trochlear nerve. Please note her condition is unrelated to MD or BPPV.
72   "Craniosacral Therapy," Wikipedia (W http://en.widipedia.org/wiki/Craniosacral_therapy)

about his own experiences with MD,[73] the pressure is supposed to squeeze out the superfluous endolymph. The patient is directed to use the Meniett about three times a day for short periods to time. Dr. Cheung found it helped some with vertigo and nausea, but not hearing, tinnitus, or balance. The one person in my study who tried it did not find that it helped. I tried a modified version of it – not called a Meniett - which did not require a tympanostomy, instead using only earplugs. The device I had was not electronic, but mechanical – you pressed with your thumb on a rubbery button. This exerted pressure on the eardrum, but probably not much of this was transmitted to the inner ear. It had no discernible effect. The Meniett Device is in wider use in Europe, where it is considered a more mainstream therapy than it is in the US. Here, it is expensive and may not be covered by insurance, although it may offer a money back guarantee.[74] I include it here because it is only slightly invasive, not risky except for possible infection associated with the puncturing of the eardrum, and some have found it useful. In general, it can be mystifying as to why people in other countries may see a procedure or treatment as helpful – or not – when it is not the case here in the USA.

Acupuncture

The couple of people whom I interviewed who tried acupuncture had not thought it made a difference, but I want to mention it here, because others with MD have tried it and found it helpful.

Curious Side Effect

"Curiouser and curiouser," said Alice, in <u>Alice in Wonderland</u>. So many inexplicable things happen, continually showing us how complicated our bodies are, and reminding us that treatment or cure for any ailment may come from an unexpected source. Russell, with a diagnosis of MD, had a brief and completely unanticipated respite after

---

73   Dr. Ed, Meniett Device & Tympanostomy Tube, www.youtube.com/watch?v=hj-vBFcTKcs
74   Web Site: Meniett Device for Meniere's Disease From Medtronic

undergoing a CT scan for sinus problems. Prior to the scan, he had a 40% hearing loss in his affected ear - a degree of deafness that was noticeable to him - and nearly constant tinnitus. Immediately following the scan - like throwing a switch - and for ten days thereafter, both the deafness and tinnitus were gone! His symptoms returned gradually over several days. He speculated that OptiRay 320, the kind of dye injected for the scan, might have done the trick. But at the moment, this remains a mystery – like so much associated with the diseases and conditions that include vertigo.

Betahistine

Betahistine – or Serc (a brand name) - isn't really a fringe treatment, nor a last ditch one. It's not necessarily the sort of thing you might turn to only when all else has failed. In Europe, it's considered a standard treatment. If you scroll through the Web, you're likely to find references to it, although it doesn't seem to be recommended often in America. It's a substance defined as "an anti-vertigo drug."[75] By those who endorse it, it's thought to dilate blood vessels in the inner ear, and relieve pressure there. I found a couple of papers that stated that a sufficient number of studies had proved the drug to be helpful.[76] In those articles, it was not thought to be a "cure" as such, nor was it thought to affect the course of the disease, but it *was* thought of as an important source for symptom relief. Dr. Hain's dizziness-and-balance website,[77] however, stated that studies of betahistine were conflicting, their conclusion being that it was essentially an "inert" substance. According to them, it had FDA approval only briefly 40 years ago, which was withdrawn after further research found it to be ineffective. It's not something your doctor is likely to prescribe here, but it is not illegal, so it can be obtained from a "compounding pharmacy" with a prescription. It can also be ordered from Canadian mail order pharmacies.

75 Wikipedia, "Betahistine"

76 Lacour, M. et al, "Betahistine in the Treatment of Menière's Disease," Neuropsychiatric Diseases and Treatment, 2007, Aug. 3 (4) and "Betahistine: A Cure for Menière's Disease?" www.thecompounder. com/alternative-treatments/specialty-compounds/betahistine

77 www.dizziness-and-balance.com

This chapter focused primarily on gentamicin and anti-viral medication as controversial treatments that made a difference for a very small number of people among those whom I interviewed. Neither substance has broad support within the otolaryngological community. For gentamicin, this is probably due to the fact it is "destructive" – it permanently destroys the function of the semicircular canals in the affected ear, along with carrying the risk of destroying the auditory function as well. Those risks have to be balanced against the possible benefits of an improved quality of life – a life without vertigo. For anti-viral medication, the lack of support seems to be due to lack of conviction about its efficacy.

None of the other treatments discussed here carry much risk – they simply don't have much support within standard medicine, at least in the US. As far as I could learn, none have been subjected to intensive clinical trials of the sort needed to establish a treatment as reliable. Regarding this, Michael Strupp and Thomas Brandt (of Germany) comment, "It is remarkable that despite the high incidence of MD and the large number of studies published on its treatment over the last few decades, there are still only very few state-of-the-art prospective, placebo-controlled, double-blind trials."[78]

---

78   Strupp, M & Brandt, T. "Current Treatment of Vestibular, Oculomotor Disorders and Nystagmus." Therapeutic Advances in Neurological Disorders. Jul 2009, Vol.2, No.4, pp 223-239/

13

Non-Medical Approaches to Treatment

Search for Effective Treatment

For any illness for which there is no clearly definitive treatment, you will find numerous claims for the amazing results of special vitamins, diets, exercise regimens, and so on. From blood letting to snake oil to a variety of present day substances that have not been subjected to the rigors of scientific study, all kinds of untested treatments have their devotees. To whatever extent a substance or procedure helps, without its being checked out on a good-sized population, it's not possible to know for sure whether a truly effective remedy has been found, or whether it's just due to a quirky individual response or a placebo effect.

What follows in this chapter is a review of two dietary approaches to treating MD, one espoused by the individual who developed it, and the other the successful result of treatment by a naturopath. Neither of these is likely to be suggested by your physician, but they are not likely to be found objectionable, either, since they don't involve ingesting potentially dangerous substances, or doing anything wild and crazy. In general, most conventionally trained physicians in the United States have little interest in disciplines such as holistic or naturopathic medicine. This doesn't necessarily mean that they actively disapprove, as long as the patient isn't taking a known medical risk.

John of Ohio's Regimen

There are many who strongly believe in the value of dietary supplements to treat vertigo. In particular, there's a website by someone who calls himself "John of Ohio," who was diagnosed with MD in 1995.[79] He presents a carefully thought out regimen he developed after considerable research. Since he originally posted it (before 2005), more than 100 people have given him feedback on it. He says that 88% report that it's been successful for them. "Successful" means significant symptom reduction, not a complete cure. John himself says the regimen eliminated all the symptoms that had dominated his life, but states it's important to understand that the disease still lurks. Thus, he recommends taking the assortment of pills for the rest of your life, in order to keep it at bay.

John's research included investigating approaches used in Europe as well as the US. He acknowledges a lack of medical acceptance, as well as little medical interest for his regimen. He takes eight pills every day. Two of them, lemon bioflavonoids and vinpocetine, are said to dilate capillaries and hence may improve blood flow within the inner ear. Two or three more are said to combat the herpes virus, since John subscribes to the idea that a viral infection may cause or trigger MD. Others are said to boost the immune system. He notes that a homeopathic preparation called "Vertigoheel" seems helpful, but that a prescription is needed to purchase it. Apparently something called "Cocculus Composition" is the same thing and available over the counter.[80] Except for that, I believe that all of the substances he recommends are available without a prescription, and may be easiest to find in health food stores. I won't go into more detail about his

---

79   Simply google "john of ohio diet."
80   Long before reading John of Ohio's website, I had once thought about trying Cocculus Composition, which I found in a local drugstore. After reading on the label the remarkable number of maladies it purported to help or cure, I felt suspicious of it and decided against trying it.

ideas here, since they are presented in a clear and organized fashion on his website.

Another Successful Diet

The following is the story of a man I interviewed who radically changed his diet, with amazingly good results. At age 47, Fred experienced an episode of acute "wooziness," leading to vomiting and very unsteady walking that lasted for hours. At first he attributed it to food poisoning, but it began happening once a week, then twice. His worst episode occurred about two months after the first, beginning while he was sitting on his front porch, when he became so dizzy he couldn't even sit up. The world was spinning and he threw up repeatedly for 21 hours. Around that time he went to see a doctor, fearful that something was wrong with his brain. He found the encounter very distressing. An MRI and a CT scan were done – in order to rule out tumors, he was told. "This was *not* reassuring!" Other, less devastating, possibilities were not mentioned. He noted that he already had a hearing loss – it had been compromised since his twenties – so he used the evaluation as an opportunity to get his hearing tested. When I inquired about tinnitus, he said he'd had it for quite a while before vertigo symptoms ever developed. It did become louder with the onset of vertigo – over time becoming so loud a roar that it prevented him from hearing much else. Other than being told he didn't have any tumors, he doesn't remember receiving any explanation about what could be wrong. He apparently was diagnosed as having MD, but thought this was decided on by default. The recommendation of a low salt diet helped a bit, but overall, he did not find his contacts with doctors helpful at all. It would have made a difference to him, and reduced his anxiety a lot, to have had a doctor who seemed to care and who took time to educate him about what was going on. He still wishes he'd been told right away about the possible "non-life threatening causes" of vertigo.

Over the ensuing 17 years, Fred continued to have periodic attacks of vertigo, but the most prevalent symptom was ongoing "atrocious"

balance. He was never able to walk more than a few feet without stepping out of line. He'd often have to hold on to furniture or the wall
as he walked through the house. People frequently assumed he was
drunk. It should be noted that he was able to continue working as an
artist despite this. He found that practicing yoga every day was valuable in that it helped him feel less upset about his impairments, and
he'd grown more accepting of his condition over time. The tinnitus
dialed back after a few years and eventually subsided to a light tone.

Things had changed a few months before I spoke with him. He'd
been having trouble sleeping, as well as with fatigue. He consulted a
naturopathic physician, who did an evaluation and then recommended
a special diet. He was to eat no grains, no added sugar. He was found
to be allergic to dairy, beef, and lamb, and was to avoid eating these
as well. Within a week of following this regimen, he was breathing
more easily, and most amazingly, his balance returned – "a thousand
per cent!" as he put it.

Fred's case is a complicated one. Both tinnitus and some hearing
loss existed for years prior to the onset of vertigo and balance problems.
Balance troubled him more than vertigo itself. As with many others,
his story is not that of typical MD, so some might believe another diagnosis to be more fitting. While it's unclear from a theoretical point of
view why this particular diet should make such a remarkable difference
in Fred's symptoms, it is consistent with the fact that dietary restrictions, as we know, are among the primary recommendations given to
patients with MD. Thinking in terms of the possibility that allergies
may provoke vertigo, it is not uncommon for people with MD to be
given antihistamines. I took one daily for two years, based only on the
thought that since my attacks seem somewhat seasonal, perhaps they
were exacerbated by an allergy. In that sense, eliminating an actual al-
lergen (in Fred's case, the meat and dairy) would make sense.

14

Brain-based Diagnoses: Experiences

This book focuses primarily on the vertiginous conditions that *arise* in the inner ear, but there are many people who have similar symptoms whose origins are thought to be within the brain. At first blush, it sounds worse to have something in your brain that's causing vertigo, and of course that might be the case if the cause were an inoperable or cancerous tumor. However, some of the people in my sample who were given brain-based diagnoses were ultimately able to find therapies that made a real difference. I thought this would be potentially useful information, since so many people spend years suffering with vertigo, finding no clarity concerning their diagnosis, receiving little or no meaningful treatment, no understanding about what is wrong. Knowing that unconventional therapies sometimes do work does not mean everyone should rush out to try them, but does mean keeping an open mind can be important.

Sometime in her 20s, Nancy woke up in the morning so dizzy she couldn't get out of bed. She assumed it was a virus, and tried to sleep for the next four hours as it wore off. She felt very sick, with a "fuzzy head" and nausea for the rest of that day, but was normal the following day. For about the next ten years, this happened about once a month, but she didn't contact a doctor, since overall she was managing all right. She didn't want anything to interfere with her life of teaching and coaching, raising her children, going skiing and playing tennis, and generally being active and enjoying the company of others. However, in her 30s the symptoms became more frequent and began to seem "relentless," so she finally went to a major medical center where she was seen by a neurologist. A CT scan and EEG were administered that showed some "scar tissue" but nothing that clarified what was going on. Presumably due to the absence of identifiable pathology, she was told the symptoms were psychological. Furious, she yelled at him and left the room, saying, "I'll live with this!" There were no further consultations with doctors for many years.

In her 40s, things got worse. The vertigo increased. She might wake up with a spinning head, vomiting, and feel horrible all day or she might get up feeling fine, then a sudden violent whirl would hit her, such that she'd need to grab something to avoid keeling over. This might happen several times a day, and she dreaded it, never knowing when it might occur. "It wore me down emotionally." During times when she didn't necessarily even have vertigo, she developed strange perceptions that led to injuries. She was prone to feeling that things around her were moving. If anything near her actually was moving, it might appear to her as if it were going faster than it was – presenting a tense and difficult problem for her while driving, since other cars are always zipping by. Once, standing near her parked car after opening the door, another car drove by, which made her feel as if her own car was rolling. Impulsively reaching out to grab it, she lost her balance, and broke her wrist. On another occasion, she was standing on a chair posting things on a bulletin board, which looked to her as if it had begun to move. In reaching out to steady it, she stepped into the air, fell off the chair and broke a toe.

At that point, it seemed like a good idea to see a doctor again. She went to an ENT and again a neurologist. Because she had no deafness and no tinnitus, she was told her attacks didn't fit the diagnosis of MD. Her first diagnosis was BPPV. The Epley Maneuver was administered, which - she said with emphasis - did *nothing*. PT was recommended, and she did go for two years, but saw no benefit. She consulted with a doctor described as a specialist in vestibular diseases, who thought she had a vestibular inflammation. She found his tests so horrible she couldn't complete them, and panicked. She quoted him as saying, "I can't do anything for you." She left crying, wanting to kill herself, thinking she couldn't continue to live like this.

In 2005, right around the age of 50, after more than two decades of all this, she suffered a catastrophic injury. While at work, she had a burst of vertigo and grabbed a freestanding bookcase nearby. It tipped over and she fell back, with the bookcase crashing down on top of her. She sustained a serious brain injury, and spent six months in a Boston hospital. She had severe problems with memory, concentration, and said her IQ dropped to 80. She was paralyzed: she would think of walking, but her legs didn't move. She was told she'd never walk again (and responded with a "go to hell" to the doctor.) She went home in a wheelchair, with physical therapists, occupational therapists, and speech therapists coming to the house. The vertigo continued. She was deeply despondent, gave up on religion, felt angry with the world. She decided to see a psychotherapist, and did find someone she liked a lot.

In March of 2006 something remarkable happened. A woman Nancy saw at church told her about a kind of biofeedback called advanced neurotherapy, practiced by Jolene Ross, PhD. She followed through immediately. She was evaluated by Dr. Ross, who administered a "quantitative EEG" that develops a "brain map" while the patient performs a series of tasks. The biofeedback therapy itself is targeted at areas identified as trouble spots on the map. Once the problem areas were identified, she went for treatment three times a week for about four months.[81] By July 2006, the vertigo was gone! She continued the treatment weekly for the next two years, stopping in 2009. Since 2006, the only time she has vertigo now is during the first day of a cold, and then only during that one day. It took about four months to get rid of

---

81　During therapy, the brain is being "retrained." The learning does not require conscious effort on the part of the patient, who is comfortably seated in front of a computer screen. There is no pain or discomfort. The average person will need about 40 sessions in all, but will begin to feel improvement after about 10. Once therapy is complete, the results last, as three and ten year follow-up studies have shown. For more information, Dr. Ross has a particularly well designed web-site: www.retrainyour-brain.com/ but remember, this is about the brain, and not the inner ear.

the vestibular symptoms that had plagued her life for 25 years, causing anguish and despair, and in the end, a dreadful injury. Dr. Ross emphasizes that she works with the brain, not the vestibular system.

It should be noted that, during all those 25 years of affliction, Nancy had been able to continue with her job as a teacher, but the traumatic brain injury made this far more difficult. She did regain the ability to walk, but had lingering cognitive symptoms that interfered with her functioning. She tried returning to teaching, but found it overwhelming. She worked in the front office of the school for a while, but it was difficult for her to keep track of things. In addition to that, a family emergency led her to decide to stay home. She has ended up teaching piano and art from her house, and in addition can ski cross-country, snowshoe, and play tennis. She's also written three books[82] - and can now say, "It's all worked out."

Some background history is important in trying to understand the course of Nancy's illness. It's pretty clear that her symptoms did *not* originate in the vestibular system. What's also clear, unfortunately, is that this was not recognized by doctors she consulted over the years. As has been seen, teasing apart the details of vertigo and its related symptoms and figuring out the underlying cause is not always easy. And years ago when she was in her 20s, if anyone had settled on the idea that an injury to the head might be the culprit, they might not have come up with a recommendation for therapy, anyway.

Nancy actually does have a history of sustaining several head injuries throughout her life. She had been knocked out several times, the first as a small child when she crashed into a bookcase. She was active in sports as a teenager, and took a major fall during a ski race, as well as getting kicked in the head playing field hockey. After the latter accident, she was unconscious for over two days. Anyone who has not responded to treatment for vertigo, especially if it doesn't have the usual accompanying symptoms, as was true for Nancy, would do well to think back over their own personal history to make sure there's no

---

82   They're available on Amazon and Barnes & Noble, by Nancy Y. Fillip. One title: *Does Spelling Count?*

forgotten head trauma. Of course, not many will have as dramatic a story as she did.

In contrast to Nancy, Cynthia assumed early on that her vertigo was a result of a head injury. About forty years ago, while tobogganing at age 25, she had an accident that delivered a blow to her forehead, resulting in a concussion, and within a year she developed headaches and vertigo. Her doctor attributed these to the head injury, but offered no therapy. She described her vertigo as "all things spinning," said she could only sleep on one side, that the other would bring on vertigo. Blinking lights or moving her eyes a certain way could set off vertigo. She never fell, but bumped into things. At first, Antivert helped to control this. Experiencing attacks of vertigo periodically, her life went on in a manageable manner for nearly 20 years. She actually sustained two more concussions during those years, with no immediate symptom exacerbation.

Well into her forties, things started to get worse, with more frequent attacks. She was checked out for MD, but that was negative. During her mid to late fifties, her symptoms became overwhelming. For two years she had vertigo nearly all the time. She could hardly eat, couldn't read at all, was afraid to drive. The Antivert was no longer effective. Her work was essentially a desk job that didn't involve much moving around, but she had to cut way back on it and work only a day a week. She felt afraid of falling, of causing a car wreck, of never getting better.

In 2006, when she was around 60, someone suggested she try Feldenkrais therapy. After four treatments the vertigo vanished – amazing after 35 years of dealing with it. There have only been two attacks in the six years since then, and those occurred in conjunction with an injury to her spine, but the therapy enabled her to put a stop to them. She also developed depth perception through the therapy, something she'd never had previously in her life. She has also learned to slow down, think about how to move through space, control her balance better.

Feldenkrais therapy does not involved specialized equipment in the way neurotherapy does. The patient is trained in greater awareness of the body and senses, how to use the head and eyes to promote balance, how to be alert to one's breathing, how to move efficiently – how to recognize your own body's feedback. The therapist uses a hands-on approach, placing the patient's body in particular positions, laying a hand on certain areas. It seems like a cousin of the craniosacral therapy (discussed in chapter 12), vestibular PT, yoga, or perhaps mindfulness. A Feldenkrais practitioner with whom I spoke said it does sometimes provide relief for those with MD.[83]

Cynthia feels strongly that alternative therapies are not all just "gobbledy gook," and clearly for her and Nancy they have provided stunning results, far superior to anything offered by conventional medicine. Neither of these treatment programs is invasive or risky, so that the main gamble you would take, if it turned out they weren't effective in your case, would be the cost. Some insurance programs might cover or pay part of the fees. However, these are the stories of only two people, and don't represent the experience of the large number that is needed for a treatment to be confirmed as generally helpful.

---

83 From a theoretical point of view, or maybe just from mine, compared to the biofeedback, it's harder to understand why Feldenkrais therapy should reduce or cure vertigo. In their website, the therapy was listed as appropriate for both closed head injuries and neurologic disorders. I was able to get information about the method by googling it, but couldn't come up with a comprehensive and well organized website.

15

Symptoms Change – My Illness

Its Third Phase: December 2008 – March 2013

By the end of the year 2008, I had been dealing with vertigo and its related symptoms since February of 2006, a matter of nearly three years.
As the end of the year approached, and similarly to the previous one, I
was guardedly hopeful that the condition was fading away. Despite periodic flashes of dizziness, zings in the head, passing pressures, plugged
ears, off balance lurches, and other things I can't even remember right
now, I'd had no symptoms that lasted for more than a few minutes
since mid-July, other than generally diminished energy. Most of the
time, I was able to do what I wanted - my daily life was not significantly impaired. I did, however, retire from my job as a school social
worker in June of 2008, a decision that was influenced by my tendency
to become fatigued.

Then, smack dab in the middle of December - and very inconve-
niently - I began to feel sick again. For the first three days, this meant
I felt woozy and queasy, but didn't really have vertigo. I was holding
onto furniture as I walked to keep from stumbling around. Moving
about made my brains feel as if they were sloshing around rather like
a thick gelatinous fluid. (During my interviews I was amused to find

that several people used the phrase "Jello brains" as I had also done.) Periodically, the nausea increased and I threw up several times, usually at night. On the fourth day of all this, I woke up in the morning with a brand new symptom. I opened my eyes and saw double! The lamp mounted up on the wall to my right was where it belonged, and alongside was a ghostly version of itself, overlapping a bit and slightly higher – or was it lower? As I looked around, everything else in the room had a twin as well, though the phenomenon was much more pronounced to the right side. I got out of bed and staggered to the bathroom, and stared at an overlapping reflection of myself in the mirror. I expected my eyes to be crossed, but I couldn't see that they were, although it was hard to look for more than a few seconds. After about half an hour or so, the doubling faded away, but the fatigue and wooziness did not. My eyes felt funny all day.

For five days in a row that December, I woke up with double vision (known as **diplopia**). Curiously, during those days the other symptoms of fatigue, nausea, and so forth began to recede. On day six, the diplopia was gone, although I experienced many quick jolts of dizziness. Through some miracle, I felt pretty well on Christmas Eve and on Christmas itself - the days when it mattered most - to be followed by feeling worse again for several days afterward. Along with the worsening, my eyes felt very peculiar, but the double vision seemed to have left.

I wasn't as concerned as I might have been about having double vision, since it arose in exactly the same way as the symptoms I'd been having for the past two years, so it seemed that it must have something to do with MD. So I wasn't really worried about a brain tumor or multiple sclerosis. Within a day or two after the diplopia appeared, I trolled through the Internet to check as to whether it was identified as a symptom associated with MD. Although I did not mark the site, I read somewhere that it *was* sometimes present in people with MD[84]

---

84   Later on, in trying to check up on this, I have not found it easy to locate sites that include diplopia as a symptom associated with MD, and don't believe I've ever read an actual discussion of the possibility. When I've seen it, it's just been placed in a dreary list of possible symptoms.

and felt reassured. Nonetheless, I made an appointment to see my internist right after Christmas, and met with him just before the turning of the year, at a time when I still did not feel well, although the double vision had left.

When I met with my internist, he thought it best that I see someone who specialized in the inner ear, and recommended that I meet again with the otologist I'd seen a year and a half before. The otologist ordered a test that was new to me, but surprisingly, I don't know what the results of it were. It seemed inconclusive – I didn't think the doctors were particularly worried, and no new treatments were suggested. I have not seen an otologist since that time, as of early 2013 (four years down the road.)

In any case, I did not feel really well until the end of January (2009), and my symptoms were so minor that I had begun to think I was imagining things, until the middle of one afternoon when I noticed my energy and focus had returned and I suddenly felt fine – and was relieved to know that the vague malaise I'd had was not just a figment of my imagination. However, throughout the winter and most of the rest of the year, I was plagued by many more days of feeling mildly affected my condition. This took the form of nausea, diminished energy, difficulty maintaining balance. I had no more double vision until mid-May, when it was present for four mornings for about half an hour, noticed as soon as I woke up. For each day that I awoke with it, the rest of the day would be negatively affected, in that I felt actively lousy and couldn't walk a straight line. After the diplopia left I would continue to feel poorly for another few days. At no time during 2009 did I have the sense that I might be recovering from MD. There were as many days with low-grade symptoms as there were normal days, so it was always in my awareness. I had two more bouts with double vision, each time experiencing it in the same way, and each time feeling distinctly unwell during those days, and for about a week afterward. Despite how significant this felt to me, other people were not apt to notice anything was wrong with me unless I told them.

For whatever impact it may have had on my condition, 2009 was a particularly tough year in my personal life. We had a series of major losses: our daughter had an ectopic pregnancy, a dear friend died of leukemia, and one of my sisters died of colon cancer. Many of the months surrounding my sister's death were deeply distressing. Visiting her often before death took her, being involved in planning the service, going to her house to help go through her possessions, all left me exhausted. We all worked together: my other sisters, her children and other family members, and we were all a support to each other, but each encounter was hard. It seemed possible that the stressfulness of that year could have contributed to the increased pervasiveness of my symptoms.

Despite my own sense of having being plagued by my illness, it is clear that being subject to diplopia has not been as overwhelming as intense vertigo, all the more so for me because it hasn't lasted for hours at a stretch. What really troubled me far more was feeling lousy so much of the time. In so saying, I'm aware that anyone who has very frequent bouts of strong vertigo would trade places with me in a heartbeat. I knew things could be so much worse, but nonetheless, feeling done in and queasy so often, with no way to know if it would ever end, was depressing. But over the course of the next two to three years, my overall health has improved. This is mostly because I feel pretty well during the times between attacks, so my whole life hasn't been suffused with fatigue. During 2010 and 2011, a pattern of sorts emerged. In January, I'd start feeling crummy for a couple of days, then diplopia would develop and I'd feel debilitated for a week or two. During most of the winter I'd do well, only to be affected again in April, May, and June. I'd do well again over the summer before getting hit with a resurgence of symptoms late in the fall.

However, over the holiday period that extended from late 2011 until early 2012, I had about two months where I felt discouraged again. During those months, I had four episodes of double vision combined, as ever, with fatigue, nausea, wooziness, and balance problems. I was sick more than I was well. But after early February of 2012, I felt normal until May, when I had a bout of double vision that lasted eight days, along with the usual accompaniments. This time, the diplopia was the most pronounced it had ever been – the twin objects in the room were less overlapped and somehow seemed brighter, more sharply defined, and more disturbing. They stayed double for an hour or more, longer than ever before. I fell twice in one night getting up to use the bathroom, and was afraid I was in for another month or two of this - or worse. And I was fearful that my summer might be ruined. Happily, that was followed by months of feeling well, months that raised my hopes, while at the same time I'd remind myself that I'd had a previous year of being essentially symptom-free from about July to December. But overall, with the exception of the two months mentioned above, things have continued to go better for me over the two most recent years. Starting a year *after* the onset of double vision, I have had fewer days of feeling really awful than I'd had previously, and longer spaces between attacks.

After my attack in May of 2012, I spoke with my internist about the possibility of prescribing anti-viral medication, and sent him a copy of Gacek's article. He agreed to do it, so for many months I carried a bottle of acyclovir with me wherever I went, so I could start taking them immediately at the onset of symptoms. In early January of 2013, I got my chance, but felt uncertain about it, because the symptoms were so mild. The double vision was subtle and soon gone each day, and the degree of fatigue and disequilibrium was slight. But I did go ahead with it, and felt discouraged at seeing no particular effect. The episode dragged on. My husband suggested the continuing minor impairment might be thanks to the medication. The whole course was scheduled to last several weeks, with the dosage dropping off gradually. Lamentably, about four days before the end of the regimen, I began feeling distinctly worse, and woke up with definite, not so subtle double vision for seven days. It was really discouraging that this attack got under way while I was still taking the pills. So I can't report, at this point, that they did me any good.

Diplopia is a symptom that arouses more concern among medical professionals than does vertigo that arises in the inner ear. Because the latter is peripheral - originating outside the nervous system - it is perceived as "benign," which I suppose means not life-threatening. Diplopia, on the other hand, tends to mean something is going wrong in the brain and/or the cranial nerves, such as a tumor, multiple sclerosis, myasthenia gravis, or a host of other conditions that could be fatal at worst, or if not that, severely affect one's life. Be that as it may, my ophthalmologist was mystified by my double vision, since it did not make sense to her that it could be part of a syndrome associated with MD. After performing several careful examinations herself, she referred me to other specialists. Over the past two years, I have met with a neurologist, an ophthalmologist who specializes in neuro-ophthalmology, and had a second MRI in early 2012, just about six years after the first one. None of this has thrown any light on the reason for the diplopia. Despite the inconclusive outcome, I appreciated the interest and curiosity on her part.

Among those whom I interviewed, there were others who experienced diplopia, but theirs may have been attributable to a condition of "central" origin, for example a result of injury to the brain, such as a previous concussion. At first, when I learned of the vestibulo-ocular reflex,[85] the connection that runs up the vestibulocochlear nerve through the brainstem and out to the eyes via the three cranial nerves that innervate the eyeball,[86] I thought that might explain how diplopia could have something to do with MD. Each of these nerves is involved in stabilizing the eyes as your head and body move around. If your head moves to the right, the vestibular nerve will send a signal to send the eyes to the left. If your head moves down, your eyes will move up. It is a reflex, which means this happens without conscious control. In addition you do have conscious control of your eyes whenever you choose, although you lose a lot of that control when you have nystagmus. So anyway, all this made me wonder whether the VOS couldn't

---

85   The VOR. It is one of the fastest reflexes in the human body.
86   Cranial nerves #3, #4, and #6, called the oculomotor, the trochlear, and the abducens, respectively. They are motor nerves, meaning the impulses move from the brain to outlying areas, controlling action, as distinct from a sensory nerve that carries information to the brain. For these three nerves, the impulses all run in one direction – toward the eyes. They don't return with sensory information to the brainstem.

direct your eyes to fixate in such a way as to form a double image if something peculiar was happening in the vestibular system, but apparently not.

I have persisted in exploring this topic because the double vision I get in the morning accompanies the all-too-familiar symptoms of malaise, wooziness, disturbances in balance, trouble concentrating, and so forth, that plagued me when I experienced the symptoms of vertigo that are characteristic of MD. To me it seems that the vertigo has simply been switched for diplopia – everything else is the same, albeit diminished in intensity. So despite the opinions of the medical community, it is hard for me to believe that the diplopia is not related to MD in my case. It has also crossed my mind that, with the variety of symptoms and varying courses of illness that can be seen even in my small sample, there's the possibility that I don't even warrant a diagnosis of MD any more.

16

Psychological Effects of Recurring Vertigo

The fundamental reason for writing this book is to bring some sense of understanding to a very disorderly disorder. Those in the throes of vertigo, or those apprehensive about when or if it will strike again, are prone to an acute awareness of chaos and unpredictability. It is not easy to convey to those who have never had vertigo what the experience is like, or how it might lead to a sense of alienation. Several people pointed out that even one attack of vertigo can be remembered vividly and with horror. The recollection often remains a strong force in one's life, and may be somewhat different from physical pain in this regard.

Physiology, anatomy, symptoms, and medical science are all important in gaining an understanding, but none of these probe the human dimension. That is best told through the experts in that domain – those with first hand knowledge of vertigo. That, of course, is one reason I chose to explore in detail the real life experiences of so many. Despite the unique character of vertigo, everything we go through is tied in some way to the totality of human experience, partly through the threads of emotion, often articulated by writers and poets, some of whom I have quoted below.

This chapter concerns people's emotional experiences when they perceive vertigo to be an ongoing affliction, or when they fear it may become so. I want to reiterate, for those who have recently developed symptoms, that not everyone with a vertigo-related condition feels dominated by it, or experiences it as chronic or life-altering. We all hope for a spontaneous recovery or for a treatment that proves to be a cure. Failing that, we hope that the intensity of the disease will diminish over time, which was the experience of the majority of those I interviewed.

Without the assurance of a full recovery, a chronic or a recurring illness can pose major challenges to an individual's sense of self and outlook on the world. No sense of certainty about ultimate recovery exists, even for those able to maintain an optimistic outlook. During periods where the individual feels reasonably well, the threat of a recurrence lurks in the shadows. Depending on the severity of symptoms and the person's makeup, some people are able to put it out of their minds most of the time, while others live with an ongoing sense of dread.

Experiencing a strong or enduring feeling of depression or anxiety is more likely to occur in people with MD, since it most cases, as we have seen, it is harder to treat than BPPV. For the most part, those who tend to have shorter, milder, or very infrequent episodes are less likely to be as deeply affected, no matter the diagnosis. However, there are many with BPPV who have gone for years without being exposed to the Epley Maneuver, and thus are among those apt to experience considerable distress. Many others with various forms of vertigo wait for a long time before any diagnosis or treatment program is offered. Suffering with vestibular problems for months or years, especially with no idea about what is wrong feels chaotic, and is frightening and deeply discouraging.

Rosamund had her first attack ten years ago, at age 35. It dragged on for days, although the first two days were the most intense. She was unable to walk without holding on, couldn't go to work, couldn't read. She was terrified, thought maybe she was having a stroke. A relative took her to see a doctor, who referred her on to an ENT who "didn't have a great bedside manner," who in turn referred her to a university hospital a two hours drive away. There she was given a diagnosis of

MD, but received no explanation. She did receive Serc, a betahistine, and the recommended dietary restrictions. She was told there was no cure, that she might keep getting such attacks, and that they would probably get less intense over time. During attacks, she felt alienated and alone. "No one understands how hard it is to function." Her employers didn't seem to believe her, and wanted a note from her doctor when she was absent from work.

When a person experiences frequent and severe vertigo, and the poor balance, trouble concentrating, strong tinnitus, or difficulty hearing that can accompany it, that person's life is going to change. This will be true whether or not they have received empathetic attention from a medical provider, and whether or not they have an understanding of their condition. Several of the people with whom I spoke had lost jobs, or changed jobs, or had to go on disability because of their illness. Even without that, when you feel deeply and chronically affected by an illness, you don't feel quite like the person you used to be. You have lost something important: your good health, and possibly your perception of yourself. When you lose something, you grieve. At the first, when it seems as if things are really going awry, you may hope fiercely that it will all pass, and find the whole thing impossible to accept. As one interviewee said, "This illness and me – we just don't go together!" As time goes along, most people do believe it more, and feel considerable sorrow, mixed with anger, about suffering through repeated episodes. This is all part of grieving, or mourning for your lost past, your lost self. What follows is some discussion of the many complicated and sometimes conflicting emotions people have experienced in trying to cope with vertigo, especially when it interferes significantly with their lives.

"Man is the only animal that blushes. Or needs to."
– Mark Twain

Twain of course tended to focus on the evils men do, on the hypocrisy of belief – the gap between what men practice and what they preach. While a person may blush for being caught in a lie or some other wrongdoing, most "blushing" occurs when one feels oneself to be an object of ridicule, or even of unwanted attention – nothing to do with moral transgression. Shaming is at the heart of human society. Embedded in our DNA, it keeps us conforming to social expectation, striving for safe status as a respected member of a group. Being found laughable or being scorned are among the most devastating experiences in life. Since preschool days, people fear being humiliated. Efforts to avoid it are threaded deeply into our characters – making mistakes in front of a group, the fear of others making fun of how you look or a personal peculiarity, or even chanting a rhyme about your name, are all dreaded during childhood. Throughout life, people continue to protect themselves from possible embarrassment, worry about "making a fool" of themselves while speaking before a group or performing, and fear being found unworthy at work or in some field of endeavor. Simply getting the facts wrong in a social conversation can be humiliating.

The sense of confidence, being able to manage, and to take charge of one's life are the converse of the feeling of humiliation. Competence and some level of personal strength are important to everyone. Mastering skills leads to this positive sense of oneself. Generally speaking, most people as they grow feel increasingly more able. But along with this, an inner fear remains of being weak, vulnerable or helpless, and consequently not respected. If life goes well, this pops up only occasionally.

There are numerous synonyms in the English language for sorrow, anger, fear, and happiness. But despite their importance, there are almost none for "embarrassed," "humiliated," or "ashamed." (I did find "abashed" when I checked a thesaurus.) Sometimes these words are not even included in lists of emotions that are used to help children learn to identify feelings. People seem to forget how strong these feelings are within the context of group life, and how much they contribute to one's anxiety from early in life. One of the features of shame is that it tends to be about something that is not within your control. It often takes you by surprise – until you heard a few snickers, you had no idea you'd spilled ketchup on your white shirt until you were already standing up to give a toast – or no idea you said the wrong thing until you noticed the other person's face change.

A person collapsed on the floor, unable to get around on his own power, needing to vomit but unable to get to a toilet, is a person reduced to a very helpless state. If this scene takes place in public, it is profoundly embarrassing, in addition to how overwhelmingly wretched or how scared the person is feeling. You are about as effective as a rag doll (and not as cheerful.) As I said to my husband when I called him during my first episode, "I'm completely non-functional." If it happened only once, you'd chalk it up as one of those weird things that catch you unawares, but when it happens repeatedly it begins to feel as if it's characteristic of you – you might think of your own self as ridiculous. You're the fool who can't walk a straight line, the fool who trips over stuff that's not even there. Beyond the fear of seeming ridiculous or foolish, or of being an object of condescending pity, lies a deeper feeling of weakness that may be hard to articulate, or that an individual may be only faintly aware of.

It's hard to look down the long road to the future and imagine being forever haunted by vertigo. Even when attacks are less severe or are fading, you are often unpleasantly affected for a period of time. You may not be able to drive, or walk very far, or walk without support. You may not be able to concentrate or read or watch TV. There may be many things – simple motions - you cannot do because they aggravate your symptoms. You may not be able to go to work, or able to do so only with great effort. You are likely to feel easily fatigued. All of this adds up to a revised sense of oneself as incompetent, to a loss of trust and confidence in oneself, and to a loss of confidence in how you are seen by others. These are related to the feeling of shame.

Many people spoke eloquently about how being helpless, and of how feeling weak and incompetent tended to make them feel they didn't amount to much. Nick, a man who had been in the navy and lived an enthusiastic and engaged life, a person with many interests, noted that he used to feel invincible, now feels vulnerable, "not whole."

"It renders you useless." Todd spoke of feeling "inadequate, dumb, not good enough." Russell was aware of a feeling that others might look down on you, see you as less capable. Jody, who described herself as a person who always valued her independence and had a masters degree in counseling, echoed the feeling of uselessness, of "feeling less than." "What purpose do I serve as a human?" Yvonne, also a professional who works as a counselor, spoke of feeling "defective, weak," as if there were "something wrong with my character." Thus, a chronic illness hacks away at one's self-image, striking at the heart of your personhood.

We all need to feel that we are capable, can do a good job, and are valued. Illness threatens that sense of strength and capability. When a doctor, or any other person, is dismissive about an illness, it promotes the feeling that you are not of any particular interest or importance.

"It is a damp, drizzly November in my soul."

– Herman Melville, Moby Dick

In addition to the humiliation tangled up in seeing oneself as incompetent, weak, and of little value, is the fact that such feelings are also inherent in depression. Even people who are usually reasonably cheerful are subject to depression when something is lost, or when life takes an unwelcome turn. They grieve for what is gone, be it a person, a job, a dream, or their own good health. Despair – the loss of hope – brings on depression. And many are filled with despair at the prospect of repeated vertigo attacks, or at the damage actually wrought by the attacks.

Sally used to live a life filled with energetic activity. She loved sports, and taught physical education in school. She had her own children as well, and enjoyed spending time with them, going places, planning fun things for weekends and vacation time. She felt at the heart of their family life. "I used to do everything and love every minute." But as vertigo became enmeshed in her life, and nothing that was done to treat it seemed to help, everything just kept getting worse. "I became a whole new me." She felt irritable, not patient at all, couldn't respond to the needs of her family. She couldn't eat, and lost 40 pounds. She was deeply aware of how upsetting this was to her children and her husband – not just the illness but her new personality - and felt powerless to change. "I felt no interest in this disease – I hated it. I felt ready for suicide."

Like Sally, Lila found the time when she was most affected by vertigo to be "the darkest place I've ever been in my whole life." From thinking of herself as a "wonder mom" who was up for anything, the person who took care of everything, she went to being the person who did nothing. Taking a shower seemed like a big deal; she hardly ever did the laundry. Cooking was a huge effort – she burned things, it was hard to get the food all organized and on the table. She lost her sense of humor. She withdrew and didn't want anyone else around. She felt "robbed" of her true self.[87]

Finding simple tasks overwhelming, losing appetite, finding no pleasure in daily life, lacking ambition, being filled with despair – these are all symptoms of a depression. Not everyone gets depressed to this extent, but many people who struggle with vertigo regularly do go through periods of depression, characterized by sorrow and anger. Most feel themselves to be diminished in some way. Many also feel lonely, or alone – isolated by their symptoms from ordinary daily life, not well understood by caretakers, a mystery to their loved ones. Quite a few subjects mentioned an increased sense of vulnerability, with some tying it into feelings about aging. Even some rather young people - in their forties - spoke about vertigo as the harbinger of things to come: a cascade of deterioration begun. People in their sixties, already aware of encroaching physical limitations, found that MD or BPPV magnified feelings of debilitation and weakness. Whenever I am laid low by my own relatively minor episodes of fatigue, disequilibrium, and so forth, I am laid low in my mood as well. Ordinary daily tasks seem like too much work, and it's hard to think of much that I'm interested in doing.

---

87   Both Sally and Lila eventually found a treatment that set their lives back on course. These have been described elsewhere. Sally was helped by gentimicin applied through the Silverstein MicroWick, and Lila by the anti-viral medication Valtrex.

It's a shame to be miserable with a depression, or depressive feelings, on top of being miserable with vertigo. A number of the people with whom I spoke had talked to their doctors about it, and had found it helpful to take anti-depressant medication, usually the "SSRIs" (selective serotonin reuptake inhibitors), such as Zoloft, Paxil, or Prozac.

> *"The six hours of deadly terror which I then endured have broken me up body and soul. You suppose me a very old man - but I am not. It took less than a single day to change these hairs from a jetty black to white, to weaken my limbs, and to unstring my nerves, so that I tremble at the least exertion, and am frightened at a shadow."*

> *– Edgar Allen Poe, "A Descent into the Maelstrom"*

In most cases, the initial attack of vertigo sets the stage for both anxiety and depression. People are stunned by the experience of becoming almost completely helpless, unable to do anything for themselves, along with the sickening intensity of the whirling, churning world. Nothing is familiar. You can't see straight - objects move around as if in a science fiction movie. If tinnitus is strong, you hear racket from within your head. The racing heart, the rapid breathing, the sweating - all are part of the body's autonomic "fight or flight" response that signals to you that you are in danger. Like the narrator in Poe's story, who escaped from the depths of the enormous whirlpool, you are shocked by your own descent into a kind of maelstrom.

More than any other emotion, people report being frightened by vertigo, especially at first, when many have no idea what is happening and think right away of a catastrophic illness. People often describe their attacks with remembered horror, the fear of being in the grip of this unknown force, the fear of what could be wrong, what might possibly happen. The fear of another attack – just the experience itself - looms large.

For most, intense anxiety quiets down after learning that the condition is not apt to be fatal, but an underlying anxiety and worry remains. This lower level of anxiety has to do with specific fears: of recurrences – the dread of ever having to endure another attack, and of the impact on one's life if recurrences are frequent or severe. There is the fear of being limited, of never being quite the person you once were, of being thought incompetent, the fear of shame discussed above. More concretely, there is the fear of being hurt in a fall, if the vertigo has a sudden and intense onset. In fact, a few people with whom I spoke actually *had* been injured, some with broken bones and others with concussions. Concern about deafness is a strong concern for many with MD. There is always the ongoing worry that a vertigo attack will strike when a special event or a vacation is planned. I've noted elsewhere the irony in the fact that being subject to vertigo is itself a source of stress, at the same time patients are advised that cutting down on stress can be therapeutic.

Underlying anxiety can be easily roused by just hints of wooziness. Once you feel off balance and the regular ordinary workaday world looks strange to you, you feel anxious. Driving, especially along a highway, is a very common worry for anyone during times when vertigo seems likely. No one mentioned actually getting into an accident, but several people I interviewed mentioned having had to pull over to side of the road to wait until the dizziness passed, or needing to call someone to come get them. One day, when I'd been feeling woozy and unwell, but did not have true vertigo, I was driving along an interstate highway. The fact that there was no chance of getting off the road except at an exit, and that cars and giant trailer trucks were rushing by me, tended to make me feel trapped and jacked up my anxiety a little. I sat forward and tensely clutched the wheel. A weird and foolish thought slid into my head. My hands might suddenly let go of the wheel, float up in the air, and I would be powerless to stop them. I would have no control over them nor over the car. It sounds absurd that I might have taken this seriously. Of course I knew I was not going to let go. But at that moment, it did seem like a possibility. I tightened my grip on the wheel and stared fixedly at the road in front of me, making sure my head did not move. I'd been worried anyway about driving for fear of an attack of vertigo, and so many unnerving things had occurred, so many episodes of loss of control, that anxiety got the best of me at that moment.

Some people with vertigo develop panic attacks, also known as acute anxiety attacks. This is a more severe form of anxiety that can take you by surprise, not unlike the vertigo itself. It is not always accompanied by a thought, meaning that the person may not be wondering if they are about to die or have morbid ideas about what is wrong, and may not even be imagining anything in particular. Sometimes it's just a wordless and overwhelming feeling of dread or danger, an I-have-to-get-out-of-here feeling. It can be so intense that people who are subject to them dread them as much as they do an attack of vertigo.

Russell first developed vertigo at age 29, nearly 20 years ago. He was by nature an energetic person, an organizer, full of enthusiasms, but once the attacks began, he could "never do anything without thinking about where safety is, what's the shortest way home." He never left the house without Valium and Compazine in his pocket, "lived in fear" of another attack. He felt generally tense, short-tempered. Luckily his job, only ten minutes away, offered flexibility, and he was able to continue to work. When he began experiencing anxiety attacks, he didn't realize what they were. He had thought, "Only weak people get anxiety." He began to notice the reaction occurred when he felt trapped in some way, or stuck in circumstances beyond his control.

He was sitting in a doctor's office, waiting for an unconscionably long time to be seen. As the minutes ticked on, he felt restless, worried about getting out in time to pick up his kids, caught between the desire to get up and leave versus continuing to wait, never knowing when he'd finally be called in. He felt as if the doctor had control of his life. His heart began pounding, he felt increasingly "wired," had a racing sensation in his head and a general fear that something was wrong, that he needed to get home. On another occasion, he was scheduled to fly to Australia on business, flying first from the East Coast to Los Angeles, where he was to change for a non-stop flight to Sydney. On the first leg of the journey, he felt deeply anxious most of the way, thinking repeatedly about being trapped in the airplane for the fourteen-hour journey to Down Under. During the layover, he sat in the airport brooding about it and remembers some guy walking past and commenting, "Somebody's not having a good day." In the end, he could not bring himself to board that airplane and had to return home. At that time, he was still "clueless" about what an anxiety attack was, and had no way to understand what was happening to him. Understanding is generally helpful in coping with mysterious symptoms – it helps to know what you are dealing with, even if you don't like it. This seems equally true for attacks of anxiety as it does for vertigo.

Matthew, a firefighter, a person normally prepared to face danger, a go-getter, is athletically fit, a cyclist. In addition to his enjoyment of physical activity, he sings bass in more than one choir. Originally diagnosed with BPPV twenty years ago, his symptoms evolved into MD more recently. Some time after a series of particularly brutal attacks of vertigo, he began experiencing anxiety attacks at random times, not directly related to bouts of vertigo. During them, he was afraid to leave the house, bedeviled with fear of an unknown danger. In some ways he finds the anxiety attacks cause more trouble than the vertigo – he doesn't want his life constricted by fears of being out in the world. He's working at trying to be "in the moment" – to enjoy life as he finds it, and to make the most of the symptom-free times.

The anxiety attacks described above may not appear to have much to do with vertigo, since they can occur at times when no vertigo is present, and are not even triggered by thoughts of vertigo. However, neither of these men ever had an anxiety attack before they developed MD. The original stimulus for anxiety was the vertigo attacks and their effect on their daily lives, followed by anxiety spilling over into other areas. As Russell said, "People don't understand how serious MD is." Thus, people who have had vertigo attacks may find themselves prone to occasional bouts of unexplained anxiety. Those who are troubled by anxiety attacks, or strong anxiety specifically related to vertigo might want to consult with their doctor about anti-anxiety medication. They also might find psychotherapy helpful. Cognitive behavior therapy (CBT) is often effective in treating anxiety.

For the sake of clarity, I presented depression and anxiety separately in this chapter. However, more often than not, they go hand in hand. You will not often find depression without accompanying anxiety. Anxiety is somewhat more apt to stand alone, but a strong dose of it frequently leads to depression. Clearly, most people with recurrent vertigo have a lot to come to terms with in addition to the vertigo itself. It's harder to think of oneself as independent, energetic, adventuresome, or just plain old capable in the face of a vertigo onslaught.

"When angry count to four. When very angry, swear."

– Mark Twain

The rageful kind of anger Twain was probably talking about is not the sort of anger primarily associated with a person's struggles with vertigo attacks. The subject of anger has come up frequently throughout this book, but it is the feeling of frustration, or a quiet and resentful sort of anger. Nearly everyone feels at least a little angry if they suffer recurrent vertigo episodes or if vertigo makes inroads into their daily life. This is not surprising - people expect to be mad when things go wrong, even if it's just a stubbed toe. And many have felt resentful if they have not experienced doctors or other caretakers as interested and supportive, or angry if the recommended treatment hasn't helped. Some are likely to feel more angry than depressed, but anger does tend to accompany depression. It has been said that depression is anger turned inward. Anger tends to seek an object, but with an illness there's really no one to blame but the hand of fate.

Anger can intrude into daily life in the form of reduced patience. When you're feeling lousy for days on end, you're likely to feel grouchy and have little toleration for mundane nuisances. Little irritations that might normally be passed over lightly loom large. Minor frustrations in getting through the daily routine, such as a drawer you have to yank open or a dropped pencil can be ridiculously infuriating. And on top of that are annoyances engendered by the vertigo and imbalance: stumbling into furniture or staggering while carrying grocery bags.

People are born with a certain temperaments, so that some take adverse events much harder than others. Some are by nature more confident, some shy, some worry a lot, some rush out to greet the world, some hang back. People are going to process vertigo differently. Whether depression, anxiety, anger, or embarrassment is at the forefront will depend upon your makeup. Just as with everything else, your core temperament makes a difference in the way you cope. The people I interviewed varied in what they had to say about their reactions to their illness. Many felt enormous distress over it, but others would say they'd learned to live with it or couldn't allow it to dominate their lives. There were those who, even with repeated bouts of vertigo, did not think of themselves as having a chronic illness. Most people eventually do make an adjustment, but all are likely to struggle with emotional reactions along the way.

17

Role of the Medical Profession

Sickness is a part of the human experience, even for those who are rarely ill. Sicknesses include exhaustion, lack of ambition and interest, weakness, pain, discomfort, loss of appetite. In most situations, people fear pain the most of these, except for times when there is the possibility that death or disability may be the outcome. Medical practitioners vary in how much interest they demonstrate in the subjective experience of illness, focusing primarily on procedures that can alleviate or cure symptoms, or save lives. As noted elsewhere, strong vertigo is not a common part of illness symptomology, and its very uniqueness often makes it harder for caretakers to empathize with the experience, and also makes it more likely to be a significant source of distress to the patient.

For the person with vertigo, his or her interactions with others make a difference in how strongly feelings of anxiety, depression, and anger are experienced - how deeply felt the despair associated with it is. When the family offers support, when people at work understand that the individual is not using or faking symptoms to back out of responsibilities, things go better. The person feels less lonely and isolated in her experience. The same is true for the patient's relationships with the physicians, nurses, physical therapists, and others who may be involved in their care. Interactions with others affect mood.

Expectations of caretakers are often high, and if they come up short, there is disappointment.

In reading this book, it will be clear that a significant number of people lived for a long time with vertigo without being provided with much understanding about what was going on. In addition to that, many went on for years without receiving treatment that made much difference, including those with BPPV. Although the sample I interviewed may seem too small to be representative, it is worth noting that I did not screen people in any way, meaning I wasn't looking for anything specific – just their histories. It was surprising to me that so many of them had had such difficult experiences in seeking help. Many felt the medical specialists with whom they met had very little interest in their situation. They spoke of feeling not listened to, being of no interest, dismissed, disrespected, lonely with their symptoms, angry. Anger toward the medical profession can occur simply because medicine may have little to offer in an intractable disease. But the feeling of being brushed off goes beyond that, and has to do with the demeanor of the doctor in meeting with the patient. People need a chance to describe fully what is wrong, and to feel the doctor is interested and is trying to understand. The sense that a doctor makes assumptions and wants to reach a swift conclusion feels dismissive, and can be intimidating. In my own case, the fact that doctors seemed to expect me to have vertigo for a few hours and feel fine thereafter prevented me from explaining the way symptoms dragged on for days. The sense of not being understood (and of not having the expectable symptoms) increased my anxiety considerably, as was true for many others.

Those in the medical profession generally are most comfortable when they can offer a plan of action that is likely to alleviate symptoms, or better yet, cure what ails you. They may underestimate the value of simply listening. Several people told me that just relaying their story to me made them feel better. Of course, nearly all medical personnel have extremely busy schedules and find it hard to allocate time for listening to a detailed story, unless they see it as leading to a concrete outcome. Yet the caring and interest conveyed by listening can be an important part of helping reduce a patient's anxiety, sense of isolation, and despair. One person in my study, who had had both cancer and BPPV, noted that she experienced more support during treatment for cancer than she had for BPPV. Some perceived the doctor as considering vertigo relatively unimportant since it isn't a fatal disease, failing to recognize the inroads it makes on one's life. As Arthur Kleinman says in his book, The Illness Narrative, "Effective clinical care …, one whose value is all too easy to underrate, is to affirm the patient's experience of illness."

One example of a patient feeling brushed off and of no interest to the doctor is the case of Cassie. She had been significantly troubled with many episodes of vertigo for three years and struggled to maintain her performance at work. Finally, she was referred to an ENT. "That doctor acted like I was the biggest imposition of her day." Cassie had noticed some hearing loss, but thought the doctor "seemed annoyed" about conducting a hearing test, surprising since such a test is standard for diagnosing MD. However, some Valium and a "water pill" were prescribed, followed four days later by a "horrible" attack. She returned to the doctor, who prescribed Meclizine and told her, "There's nothing more I can do. You're just going to have to live with it," and went on to say that the hearing loss was "mostly in your head." This seemed equivalent to the doctor's saying, "Don't bother me with your silly complaints." Very few people with whom I spoke were treated quite this abruptly and unsympathetically, but nonetheless many had the impression vestibular illnesses were not seen as of much concern.

Thinking of vertigo as a minor condition in the great panoply of diseases seems ironic to me, since MD in its worst form, as well as lingering symptoms from VN, and sometimes even BPPV, compromise a person's quality of life. An illness that makes you desperately ill several times a year, or even just feeling really crummy a good part of the time, an illness that makes it difficult or impossible for you to continue with your career or keep on working, an illness that makes you stagger around like a drunken sailor, an illness that prevents you from doing things you'd once enjoyed, an illness that robs you of your hearing, is not a minor illness. A person who is gravely affected by vertigo, or even one who just has occasional sickening attacks, needs to feel this acknowledged by the professionals he turns to for help.

Delay in finding a medical practitioner who provides a diagnosis and can help the patient understand what is going on is also a common source of heightened anxiety and despair. It appears that there are in fact many doctors who are not familiar with vestibular disorders, and another group whose understanding of them is not extensive. This is partly why several people commented to me about the importance of being your own advocate. Familiar or not, medical personnel may well find it frustrating to treat a person with vertigo, especially in the case of MD. After suggesting that a low salt diet (along with the other dietary restrictions) is the best treatment for MD, there is the inevitable disappointment if it doesn't help much. Physicians need to make it easy for patients to give them feedback, to review with them the pros and cons of other options, and be willing to explain the circumstances under which more extreme measures might be contemplated. Too many people are told there's nothing more to be done, or are discouraged from pursuing more aggressive or unorthodox treatment. Feeling that the doctor is interested in working with you to find help, as well as acknowledging to you that there is no magic bullet treatment for MD, would be experienced as supportive by most. In cases where the diagnosis seems uncertain, it would be helpful for the patient to know that it may take a while, but the doctor will be hanging in there with you while you wait to see how things develop.

Some parts of this book, including the passages above, may read like a critical attack on the way many doctors have dealt with patients who present with vertigo. This does not mean that physicians have been willfully callous or indifferent – more likely, many have had a limited range of experiences in dealing with vestibular illnesses. And of course there are many doctors who are very helpful to their patients, no matter the illness, and are admired and respected. Not everyone will be enthusiastic about the same doctor or nurse. It can amount to nothing more than a meshing of personalities – some you feel comfortable with, others not. But surely it is a good idea for everyone who works with sick people - those at their most vulnerable - to cultivate a greater awareness of their patients as people whose lives extend beyond their illness. Arthur Kleinman, in his book, The Illness Narratives, makes a distinction between treating illness, which he defines as having to do with an overall humane approach, versus disease, which has to do with using scientific findings to treat symptoms. As he puts it, "The essence … is captured by the words *empathic listening,* … which I take to be the craft of the clinician who treats illness, not just disease."[88] Support and interest go a long way toward helping people to cope with a distressing illness, and to find a way to get on with their lives, whatever limitations may be imposed.

Over the past several years, training in medical schools and residencies has become more focused on trying to help doctors perceive and relate to their patients in a more personal way – to see them as "whole people," with a focus that extends beyond their illnesses. Raising medical students' awareness of cultural differences and of values that may affect perceptions of illness, so as to enable them to try to see an illness through different eyes, is part of this effort. Part of the idea behind this is to prevent patients from feeling dehumanized in their contacts with the medical world. This approach should prove helpful to everyone struggling with difficult health issues.

---

88   p.228, op. cit.

18

Prognosis in Menière's Disease and BPPV

The word "prognosis" means foreknowledge – it means that you know or can predict something that will happen in the future. Prediction, of course, is almost always uncertain, and is based on one's past experience or what is generally known to occur in a given situation. The word is primarily used with regard to illness, and most people think of it as referring to the ultimate outcome, as in: "Will I get better?" Any time something goes wrong with you prognosis matters enormously, particularly with regard to the chances for survival.

Colds, flu, appendicitis, various infections, chicken pox, broken bones may all lay you low but the basic expectation, even for worried pessimists, is for a temporary inconvenience. For a great many of the medical conditions that strike you nowadays your prognosis is very good, although an underlying dread of incurable or potentially fatal diseases may lurk somewhere beneath the surface. Cancer or catastrophic accidents probably head the list, but people are mindful of other possibilities. In almost all circumstances, people expect a great deal of doctors, thanks to the dramatically more effective medical care achieved during the course of the 20th century. During that time, immunizations against many epidemic diseases became widespread, antibiotics were discovered and meant that fewer infections would move on to become life-threatening, chemotherapy was developed, surgery made huge advances. All kinds of tests have made it possible to gain a better understanding of what ails you and how the body works overall. 150 years ago, desperately ill people turned to doctors knowing full well that death was as much of a likelihood as was recovery. Now we tend to expect everything to be treatable, even if we know better in one side of our minds. It's hard to have a disease with no very effective treatment, one that keeps popping up – kind of like a Whack-A-Mole – making you feel miserable whenever it does. It's hard not to feel resentful that doctors are not able to do more.

Early on, it becomes clear to most people diagnosed with a vestibular illness that their disease is not fatal, so in that limited sense, their prognosis is good. But the word "prognosis" can also be used to refer to the *course* of an illness – how difficult will it be to endure it, how long it will last. People can usually face going to hellfire and back as long as they can expect to be entirely recovered upon their return. Upon learning that these attacks of vertigo tend to recur, the most important question, to me and to most others, has always been, "Is this ever going to go away completely?" It hasn't been easy to get a clear answer to that question. It's not often addressed in websites or articles about vertigo.

Before moving on to a discussion of prognosis in BPPV and MD, I want to reiterate what was discussed much earlier in the book concerning prognosis for VN and labyrinthitis. The most common outcome for both these illnesses is for a full recovery, although it may take a few weeks. They seemed to be triggered by the kind of virus that gets into your system and then runs its course. In a small percentage of cases, the infection may do some damage to the vestibular system, leaving the patient with some problems with balance or residual dizziness, or both. This is more likely to be the case with VN than with labyrinthitis. With VN, a small number of people may suffer repeated episodes of acute vertigo, so that they too find themselves coping with a recurrent illness. This was true for the two people I interviewed who were diagnosed with VN. It could be that the virus lingers on in some instances.

With regard to BPPV, the answer to the question of whether the vertigo is ever going to go away is less complicated than the answer for

MD. Starting with the least intense form of the disease, there are many with BPPV whose symptoms are pretty mild. They may feel poorly and have some vertigo for a while, but not of the magnitude that renders you unable to walk. There are others for whom the vertigo is a brief burst, lasting only a minute or two, so it doesn't dominate the day. In such cases, knowing whether or not it's going to return doesn't feel like such a desperate issue. They're apt to feel they can put up with a degree of vertigo from time to time. Ursula told me she knew how to position her head to avoid vertigo during the days she's prone to it so that trying the Epley didn't even seem worth the bother to her.

Then there are those with BPPV who have very frequent, or stronger and sometimes disabling, episodes of vertigo, clearly not the sort of thing most people can live with comfortably. For them, the Epley Maneuver usually makes all the difference. Sometimes a successful treatment actually does eliminate recurrences. The dizziness-and-balance website reports that following treatment, about one third of patients will suffer another attack after about a year, while two thirds will still be feeling fine. About half will have an attack by around five years after treatment.[89] That means that 50% are still essentially vertigo-free after five years. Overall, that same website reports that the Epley is effective in 80% of the cases. I would take "effective" to mean that it eliminates the symptoms present at the time, not necessarily that it prevents them forever. In sum, people diagnosed with BPPV can expect that the Epley Maneuver will stop a vertigo attack, with an excellent chance of avoiding a recurrence for a good long time, perhaps for always.

So the prognosis for BPPV, while not perfect, seems to be a pretty good one. But the outlook is muddled by the possibility that you might not respond well to the Epley, or by the possibility that the diagnosis may not be all that clear. Some people appear to have crossover symptoms that make their diagnosis less certain, so it may be hard to judge whether MD or a vestibular neuroma, or something more unusual could be a more likely diagnosis. Occasionally people are thought to

---

89   dizziness-and-balance.com Information provided through Chicago Dizziness and Hearing, affiliated with Northwestern University. Timothy Hain, MD, appears to be the main contributor to the site.

have both BPPV and MD. In any of these situations, predictions concerning the course of the illness are going to be more elusive.

Prognosis in MD in both senses – whether entirely shaking off the disease is a possibility, and what the course of the condition will be in the meantime – is much more variable. This uncertainty, coupled with the absence of any therapy that is markedly effective for most, poses a challenge. The fact that the symptoms come and go, regardless of what you do, reduces confidence that any particular treatment has been truly effective. Even with treatments that do seem to control symptoms, the prospect of putting up with repeated bouts for years on end can be a gloomy one, since there's nothing that works as well as the Epley does for BPPV. One bright note, as I've noted elsewhere, is that the vast majority of the people I interviewed found the earliest attacks to be the worst ones, and that most had found that either the course of the condition had dropped to a level that seemed tolerable, or had found a treatment or regimen which made a major difference, even if it didn't amount to a cure. Still, everyone wonders whether they're stuck with a chronic, albeit remitting, disease.

A helpful article was published in 2010 called "Long-term Course of Menière's Disease Revisited." It's a paper that reviews the existing literature, not one that reports on experiments or first-hand observations. The three authors, affiliated with a university in Munich, examined 46 articles about MD, covering 7852 patients all told.[90] The authors used for their study only those who'd had long-term follow-up, a reliable diagnosis, and "quantitative measures of vestibular and auditory function." On the whole, their conclusions seem somewhat reassuring. They concluded that it is likely that both the frequency of attacks, as well as their intensity, will diminish over the first 5-10 years. "Diminish," of course, is not the same thing as "disappear," although one might rightly argue there will always be some uncertainty as to whether the attacks have truly disappeared. Several people I spoke with

---

90   Huppert, Strupp, & Brandt, "Long-term Course of Menière's Disease," Acta Oto-Laryngologica, 2010; 130: 644-651. You can find this article by googling it, or you can check on Dr. Ed's wonderful YouTube site, and listen to his description of it. (See "Resources" at the end of the book.)

had gone for years without trouble, only to be hit again. This is also reported in the literature.

In reading details in the body of the text, such conclusions do not seem as clear as they sound when presented in summary. Different studies produced widely varying results. One researcher stated that the frequency of vertigo attacks remained constant over the years in 46% of subjects. Another reported a continuous decrease in symptoms over time, another found attacks had disappeared for half of his subjects within 18 years. The number of years reported on was not standard, and my guess is that patient's reports of their attacks and experiences were not consistent, either.

The incidence of other symptoms associated with MD was also reviewed in this article. Hearing loss, a nearly universal feature of MD, was likewise found to occur mainly in the first 5-10 years, then apt to stabilize. Whether this is good news or not depends on how rapidly your hearing deteriorates during those years. One of the hallmarks of deafness attributable to MD is that it fluctuates, becoming worse during an attack, then bouncing back in between. But each attack tends to make the hearing loss worse, so that the impairment becomes progressive. In my small sample, there were five people who had suffered severe hearing loss, two of whom were affected in both ears. Most of the others in the sample, including myself, did experience some hearing loss, but were not gravely affected. There were some whose hearing loss began before symptoms of MD did, thus confusing matters. Thirteen of those diagnosed either with BPPV, or who remained undiagnosed, also had a hearing loss. Most of those did not believe it to be related to their condition.

Tinnitus, as you may remember, is listed as one of the four symptoms that make up the syndrome of MD. The degree to which it interferes with life varies, depending primarily on how loud and intrusive the sound is. This is something that can fluctuate unpredictably over time, but doesn't generally seem to remain at a constant level – a good thing if you are one of the unlucky ones for whom it can be strong enough to dominate your awareness. Most of the time for most of the people with whom I spoke, it was largely a background annoyance, not something to give much attention to, except maybe in a quiet room where not much is going on. Some found it to intensify during or in conjunction with vertigo attacks, but then to moderate as the attack settled down. Remember that, like hearing loss, tinnitus is not considered to be part of the diagnosis of BPPV, but many people with BPPV do suffer with tinnitus, as do many who don't otherwise have a vestibular disorder.

Clearly, it is not easy to nail down a reliable prediction as to the course or the ultimate outcome for people with vestibular disorders. The course of the illness may be very different for people with the same diagnosis. There is the possibility that our current diagnostic labels are not sophisticated enough, that it is presently unclear how much importance to give to these differences, and whether treatment recommendations could be more effectively targeted. The most universally endorsed treatment for MD is to reduce the amount of sodium (as well as caffeine) in the diet. Yet many of those with whom I spoke – including myself – did not find that this made much difference. It's a great recommendation if it works since it's good for your overall health and has no bad side effects. Clearly everyone should try it, but in cases where it doesn't help, the question might also come up as to whether this could be attributable to a different form of MD, or possibly another syndrome altogether. Some people with MD seem to plunge down into a marked hearing loss, while others trundle along without much loss. Do those with the hearing loss have "real" MD, while others have another similar disease, or is one just a more serious version of the other?

In thinking of people in my study who were helped by the Epley Maneuver, there is a remarkable variety of symptoms, such that it's hard to think of them all as having exactly the same condition. There is Yvonne, who for years and years had extreme attacks of vertigo, including the kind that that makes you feel as if you are falling, who was initially considered to be suffering from a psychological condition. There is Frieda, whose main symptom was intense disequilibrium – she was always walking on rubbery, unstable surfaces, bumping into things as she went. The vertigo she suffered was confined to short bursts, over and done with in a matter of minutes. This also went on for years. Then there is Helen, who woke up one morning and saw the ceiling moving, and was unable to stand unaided, symptoms that are familiar to anyone who's had vertigo. She was taken to an emergency room, and received some Meclizine and the Epley Maneuver. Over the course of the following two years, she had four more such episodes, all successfully treated by the Epley.

The fact that all three of these people responded to the Epley Maneuver is terrific, and maybe it's quibbling to bring up the question of whether they all warrant the same diagnosis. But when you hear their stories, they don't sound the same.

Some people's descriptions of their experiences sound like a mixture of MD and BPPV. Probably most people with MD have some vertiginous reactions to the position of their head, particularly looking up or rolling over in bed. Some people I interviewed had an initial diagnosis of BPPV, which was later changed to MD as symptoms developed and changed. As noted earlier, there are those who believe MD and BPPV are not really two separate illnesses.

In terms of the possible long-term course of illness, there are two other unpleasant phenomena associated with MD addressed in the article that I did not discuss earlier in this work. The first is "drop attacks," some of which are known by the name of the person who first described them: "Tumarkin Otolithic Crises." During such an attack, the person suddenly falls, with no previous warning, without extended vertigo, and without loss of consciousness. Tumarkin's paper, "The Otolithic Catastrophe: A New Syndrome," was published in the British Medical Journal in 1936. Tumarkin saw the utricle and saccule as moderating posture and muscle tone,[91] and believed that during an attack, support for maintaining that tone was suddenly withdrawn, such that the individual would collapse. I pictured this as similar to a

---

91   During the time of Tumarkin, it sounded to me as if the functions of the vestibule were less well understood than they are today.

marionette flopping to the ground when the strings are released. He described the attack as taking place during a time when the person felt fine, to be followed by that person's getting back to his or her feet right away, and returning to feeling normal. (This, of course, would assume no injury was sustained in the fall.) Other writers have stated that most patients feels pushed, or "thrown to the ground" during such attacks.[92] If a person suffers repeated attacks, s/he tends to fall in the same way each time, indicating that a specific site within the vestibule is malfunctioning. Most agree that susceptibility to the attacks is associated with those who have had MD for quite a while. Sometimes a person only suffers one attack, sometimes there will be a few, then it's likely to pass. One point to be clear on is this: If you fall to the ground during an attack of vertigo, that is not the same thing as the attack described by Tumarkin. His contribution was to be a close observer and to detect this distinction.

Rick suffered several Tumarkin crises, and described them as "like a sudden drop down an elevator shaft. I couldn't raise my hands to protect myself." Despite his experiencing ten to twenty of these over a couple of years, and falling flat on his face once while walking in the woods, he had never been badly injured. Several had occurred while he was sitting at his desk, so that he just collapsed in his chair. Drop attacks are said to happen in only 1-2% of MD patients, so they are not common. If they happen at all, they are more likely to occur later on rather than in the early years, and they tend to be a phase – they don't usually last. Both things were true for Rick.

The second unpleasant phenomenon is that vertigo in MD sometimes arises in the "**contralateral**" ear. This means that vertigo could develop in the previously unaffected ear in what seemed to be an average of about 20% of cases, but this figure varies from study to study. A very few people may suffer with this from the start, but this is very uncommon (2%.) Generally, if this occurs at all, it appears after a few years. Some studies reported it as occurring after five years, others ten years or more. The possibility that this may happen is one of the factors

---

92   Lanska, D, Medlink (Neurology) clinical summary "Drop Attacks," updated May 2012

that can make physicians hesitant to provide treatments that may further compromise hearing in the affected ear.

"Burn out" is an aspect of MD that is not mentioned everywhere, but which is emphasized by some patients, and thus presumably by some medical practitioners. At first blush, it sounds like a good thing: the disease has "burned" itself out – the vertigo is gone. What makes it not a good thing is that, at least in theory, the cilia inside the semicircular canals have all been destroyed by the disease, so they no longer send *any* signals to the brain. According to the MD Information Center,[93] tinnitus, aural fullness, and progressive deafness can continue on. Remember it is expected that the other inner ear will compensate for the loss and take over the balance function. If a person's illness does progress to this point, they've probably had a pretty tough time with it down over the years. It is not the inevitable outcome of MD. In any case, the effects of "burn out" sound similar to the effects of gentamicin injections, a procedure many patients choose and have found helpful.

It remains that people diagnosed with MD face considerable uncertainty, both with regard to the course of the disease, and with regard to the ultimate outcome. It could be said that the prognosis is guardedly optimistic for most, in that the condition is likely to improve over time even if it doesn't completely disappear. This dovetails with the aggregate experience of the people I interviewed, although sadly, not for everyone.

It hasn't been easy to gather this amount of information concerning the course of these illnesses or the long-term outlook. The sense of chaos and the stressfulness of living with something unpredictable and so often overwhelming is not well acknowledged in the literature. While they may state that attacks can't be predicted and seldom give warning, there's little acknowledgment of the variations in intensity or symptoms. Websites, articles, books, and doctors tend to imply that MD is treatable, but that statement isn't so reassuring when it doesn't jibe with experience. If a low salt diet and various medications intended to control dizziness don't lead to an obvious improvement, the patient is likely

---

93   Website: MenieresInfo.com

to feel cast adrift, especially if the impact on daily life is considerable. Also problematic is the way an attack of vertigo changes perception, so you feel plunged into an alternate universe. Seeing furniture sliding around the room or feeling your cozy bed plunging down through the floor are deeply unsettling experiences. Information from your senses becomes radically unreliable. Nothing is as usual. Medical practitioners would do well to acknowledge the confusion and individual variations that are possible, rather than the focusing on average expectations for the course of the illnesses.

19

Where's the Silver Lining in All This?

It's common to hear people who have had a brush with death or a serious illness talk about gaining a greater appreciation of their lives and of those whom they love, finding new meaning in the ordinariness of daily life. After a catastrophic stroke, an aunt of mine spoke of the joy of simply looking out her window to see how beautifully green the grass was. So it occurred to me to ask my subjects whether anything good had come out of their experience with vertigo. Very few answered with a strong affirmative, although some did report positive side effects. Many people were actively negative. In that category, my favorite response to the question was, "Hell, no!!" As is surely clear by now, a great many people had a considerable amount of anger about having vertigo, sometimes coupled with anger at feeling "dismissed" by medical practitioners. However, in answer to the question, the majority of interviewees answered in a neutral tone, "I can't think of anything," or just plain, "No." It seems safe to say that few people thought of themselves as gaining inspirational or spiritual growth from having vertigo.

However, several people said they'd learned how to handle the medical establishment better, that this was a good thing. This was often in the context of speaking about not being offered much support. They spoke of needing to be their own advocate, pushing for explanations or testing or treatment that you as a patient think could be helpful. This sometimes helped them to feel less of a victim and more competent. There were a small number who talked of developing a "good attitude," thinking that you could work toward an acceptance of living with vertigo - if "this is the worst that happens, I'm home free," or "It's not so bad." Some spoke of learning to be "in the moment," such that you can actively enjoy the good times when they come, and not dwell on the hard ones. People mentioned the importance of struggling against despair, and pushing to continue enjoying your activities and achieving your goals. It's all the more important to get out and do rewarding things during the times when you're feeling normal.

One woman said it used to make her mad when people said, "Adversity helps you grow," but over time, she came to feel this has been true for her. It enlarged her perspective on the problems of others, especially of the children she works with, who struggle with limitations. "It's been a real eye opener." Several other people made similar remarks about its having increased their feelings of compassion. Some said they're more inclined to take it seriously when a friend says s/he's feeling badly. Along with this, quite a few people said they appreciated the love and support of their families when they were down. It meant a lot.

It's worth noting here that people need to develop their own thoughts about coping with vertigo or any other major life challenge. To have someone else say, "You just have to learn to live with it," or "What doesn't kill you makes you stronger," or any other platitude of that sort is not helpful. It's bound to make that person feel the condition is being trivialized, their experience disrespected and not understood.

People do find ways to cope, however, no matter the terrible challenges that threaten life as they once knew it. Some feel more positively about it than others. The story of Rick is certainly a case in point. He is an ornithologist as well as an evolutionary biologist, and is among the few I interviewed whose entire lives had to be revised because of serious changes wrought by hearing loss and vertigo. In 1988, at age 27, as part of his graduate studies of birds, he spent the summer doing fieldwork in West Africa. At that time, he had developed a unique and remarkable ability to recognize hundreds of bird songs, and previously had been able to construct a family tree of an unusual group of birds (the manakin) in South America by listening to their songs. The day before he was scheduled to fly home from Africa his ears began ringing. The next morning, he stood up and fell flat with vertigo and vomiting. This took two days to resolve, delaying his departure. He noticed an accompanying reduction in hearing in his *right* ear (a major loss which has remained unchanged.) Once back, he went to the Massachusetts Eye and Ear Infirmary and was given a diagnosis of a middle ear infection. Although his balance was restored, the hearing loss persisted. Within the month, he was diagnosed with a "sudden idiopathic hearing loss" that was probably the result of a virus.

About four years later, he began having short bouts of aural fullness in his *left* ear, along with vertigo and a low range hearing loss that was ultimately diagnosed as Menière's Disease. The disease progressed into dramatic vertigo around 1995 (eight years after the initial incident) and during the following five years he experienced two kinds of vertigo. At first these attacks were of the strong rotational sort, lasting for a few hours and happening as often as several times a week. Later on, the vertigo changed to the kind characterized by a sensation akin to a sudden drop from a precipice. The latter were short-lived and over in a minute. He was normal between attacks, but felt himself to be a burden on his family, especially when he was unable to drive. The biggest fear for Rick was the threat to his hearing, since his life's work depended on his amazing hearing acuity. He'd already had a significant loss in his right ear, and now his left ear was being damaged by the recurrent vertigo attacks. He was preoccupied with how to prevent the disease from progressing, and wondered about having surgery to cut the nerve. He tells a poignant story of taking a team of researchers, in 1998, to study the velvet asity (a bird in Madagascar), and being delighted to find one right away. As he watched it through his binoculars he was shocked to realize he could no longer hear it at all. It was "inaudible to me."

He needed to find a new direction for his career. He became an expert on feathers, and it was he who helped develop a theory of the evolution of feathers that includes the idea that dinosaurs were feathered creatures. Despite the switch in the focus of his work and the challenges to his health, he became a professor at Yale in 2004. Shortly before then, his left ear seemed to burn out. Hearing was essentially nil and vertigo attacks faded off. He has moved on with his work, studying color in feathers, and working on a theory of the role of beauty in evolution. He received a MacArthur fellowship in 2009, nicknamed the "genius grant," awarded to people involved in innovative work who show promise of continued creativity.

His accomplishments are exceptional, but a great deal has been lost to him along the way. He is now hearing disabled, often misunderstands what is said, doesn't go to movies because it's too hard to hear, can no longer learn a foreign language easily – another thing that used to be a talent. Rick has abilities that few people have, including an extraordinary focus on his interests. He did find a way to move on, but as he says, "I had a world class ability, and I lost it." He went through bouts of depression and anxiety, but was motivated to do whatever he could to maintain a rewarding and purposeful life. He reminded himself that each attack was finite – it would come to an end. He worked with a physical therapist and also found that meditation helped him to stay calm, keep his heart rate down, and control vertigo and depression. He got a hearing aid at the age of 35 when he realized he couldn't understand what his young children were saying. He tries to avoid being bitter, to avoid feeling victimized by his disease. His task is to manage it so he can keep on living meaningfully. He thinks everyone with a debilitating disease needs to do the same, but when asked if anything good has arisen from the experience, he said, "Absolutely not." Like others, he spoke appreciatively of the love and help from his family, although we'd all prefer to experience that without being hammered by illness.

20

History as It Relates to Understanding Vertigo

A Sketchy Overview of Human Development and Medical Understanding

For most of *homo sapiens'* 100,000 years or so of history, no one had the least idea what caused vertigo. There are not a lot of historical references to dizziness: I was unable to find any special ancient myths about its possible significance. Any sort of real understanding of how it works is probably not much more than 150 years old. Extreme attacks of vertigo, especially those that are recurrent, probably resemble epilepsy more than they do any other disease, in that they tend to come on suddenly and are incapacitating when they occur. In fact, in the 19th century vertigo was actually thought to be a form of epilepsy, or a start-up version of it that was likely to progress into a full-blown seizure disorder. Recently, there have been some investigators who believe that Vincent Van Gogh, famous for cutting off his ear, suffered from vertigo.

Nearly a million years ago, protohumans, the predecessors to modern man, began to develop systematic tool use and create shelters for themselves. Around 100,000 years ago, the creature we may think of as the finished product – namely ourselves, or *homo sapiens* – appeared in Africa, and began the journey that eventually led to the inhabitation of the whole world. Over the nearly unimaginable millennia since that time, people have been figuring out how to gain greater and greater control over their lives, and over the forces that affect those lives. For most of those 100,000 years, people foraged in the wild and hunted for their food. But they also designed increasingly specialized tools to help with hunting and with building shelters, and they engaged in creating beauty, in the form of cave art, jewelry, personal decoration, and music. They also fought each other and worked tirelessly on increasingly efficient ways to do that. Archeological excavations show that, long before there is any written record or civilization as such, people wondered how to understand the forces of nature, our place in the universe, and the mystery of disease. We can also see, in the pre-literate societies that existed into recent eras, for example the Native Americans, that such preoccupations and the customs surrounding them were rich and compelling.

Around 10,000 years ago, humans began to learn not only to grow crops, but how to cultivate them in order to produce new and better foods. (You need only to reflect upon the fact that you have never seen corn or tomatoes growing in nature to realize how complex this achievement was.) Agriculture led people to settle in one place, and cities began to take shape, along with increasingly intricate dwellings, social systems, and laws. These ancient civilizations developed more formalized theories about deity, about who or what controls the weather, what is responsible for the vicissitudes of life, as well as about the causes illness. All had rituals intended to influence these forces, and to honor and appease the gods who were believed to be in charge. Most attempted to heal the sick by appealing to the gods, by spiritual incantations, or by the use of remedies believed to have a favorable effect.

The ideas of ancient peoples concerning disease, regarding both cause and treatment, seem mostly superstitious to us. But along with mythical explanations, the Egyptians did study, recognize, and treat many conditions in a way that would seem familiar to us today, as did the Greeks. The early Greeks, especially as made famous by Hippocrates in the 4th century BC, began making systematic observations of disease.

In the work credited to him (the Corpus), Hippocrates identified vertigo as a symptom, and thought of it as a disease in the head caused by too much blood. Plato made direct mention of vertigo and considered it to be related to anxiety. Galen, an early Greek physician (2nd century AD) made extensive use of dissection to understand the workings of the body, and is credited with naming the inner ear the "labyrinth," a term we still use today. Over most of the 2500 years since those first complex civilizations, medicine did make gains in learning more about the structure of the body, sometimes combined with the use of helpful medicaments that were mostly discovered through trial and error. But overall, the cause of most diseases remained unknown, many treatments used had no demonstrated value and sometimes did harm, and there continued to be a marked sense of mystery surrounding them. Disease was often thought of as reflecting the will of God, sometimes seen as punishment, sometimes seen as the work of Satan. People with conditions that affected their behavior, such as psychoses, epilepsy, and most likely vertigo were sometimes viewed as "possessed."

About 200 years ago, in the early 1800s, enough had been learned and described over the past few centuries to enable an increasingly systematic study of the properties of disease, its connection to anatomy, and how to address it. However, as new findings were reported, they did not always gain immediate acceptance, as illustrated by the following brief stories: In the 1790s, Edward Jenner discovered that inoculation with cowpox could prevent smallpox, a major breakthrough, but another sixty years would pass before an understanding of germs would be widely accepted. Amazingly, despite Ignaz Semmelweis's studies[94] on the value of clean hands in caring for patients (published in 1847), it was not until sometime well after his death in 1865, when Louis Pasteur was able to demonstrate the existence of microbes, that his findings were vindicated. Even that heightened awareness did not lead to an immediate endorsement of the importance of cleanliness. Not only were they reluctant to wash, but doctors continued to operate wearing clothing covered with the blood and tissue of previous patients. History buffs may remember

---

94   There's a great novel about Semmelweis called <u>The Cry and the Covenant</u>, by Morton Thompson.

that President James Garfield, shot by Charles Guiteau in 1881, did not die as a direct result of his injuries. He died about two months later, after suffering infections caused by his physicians poking grubby hands and instruments into his wounds in an attempt to dig out a bullet.[95]

The vision of illness and infection as punishment, or as caused by possession of evil spirits, lingered on for many throughout the 19th century. And apart from that, people are creatures of habit who tend to resist new ideas and cling to old ways of doing things. Blood-letting had been in wide use for centuries, and some even remained convinced of its value until the early 20th century, with people believing they were releasing malevolent humors, or easing pressure on the heart.

At the dawn of the 19th century, many of those who thought in terms of physical, rather than mystical causation with regard to vertigo assumed it was attributable to something going wrong in the brain. Noted above is the fact that it was considered to be a mild form of epilepsy. Some thought of it as a form of hallucination. Some thought in terms of "cerebral congestion" or too much blood in the brain – not very different from Hippocrates.

Into this mix of mystery and nascent scientific thinking, two remarkable men were born in France. Both became what might be termed "Renaissance" men – people who were accomplished in their fields, but who also had wide-ranging interests, and an ability to think beyond the confines of accepted understandings. One of these men was Marie-Jean-Pierre Flourens (1794-1867), the other Prosper Menière (1799-1862). Both men studied medicine, and both were awarded the degree of doctor. Flourens, a physiologist, was a pioneer in the study of the vertebrate brain, who worked on it directly by surgically removing portions in order to see what would happen. Much of his work was done using pigeons as his specimens. It was he who established that different areas of the brain controlled specific functions. He was also a pioneer in the use of anesthesia, and presumably administered it when he prodded and dissected the brains of these birds. He was the person

---

95   A fascinating non-fiction book about Garfield and his death is <u>Destiny of the Republic</u>, by Candace Millard.

who first recognized that the function of the semicircular canals was to regulate balance and posture, and that it was another part of the inner ear (the cochlea) that regulates hearing.

Prosper Menière, who first described the condition that was named after him, was born just before the turn of the century in Angers. Angers is a city southwest of Paris, not far from Nantes, and occupies both banks of the River Maine. Several bridges span the river, connecting both sides. For centuries, the town has been dominated by a massive castle, with the unsurprising name of Château d'Angers. In this city of flourishing gardens and trades, Prosper was the third of four children born to his father, who was a well-to-do merchant. Very little seems to be known about his early life, except that he showed a great deal of promise as a student. At age 17, he began preparatory school at the university in Angers, where he won a prize for excellence two years in a row, in 1817 and 1818. He began his medical studies in 1819 at age 20, at the Hôtel Dieu Hospital in Paris. The Hôtel Dieu was situated in the heart of the city on the Île de la Cité, and was its oldest hospital, having been founded in the 7th century. The middle years of the 19th Century represented a golden age for France, when the most advanced medicine in the world was being taught and practiced there, with students coming in from many different countries, and the Hôtel Dieu was one of the most respected places to train. Menière again excelled, winning several awards, and received his "doctorate" in medicine in 1826, at age 27.

Over the next ten years, he gained medical experience during two major events. He had intensive experience in treating wounds during the bloody uprisings in Paris in the year 1832, those memorialized in Les Misérables. Later in the decade, around 1835, he worked in Aude in southern France during a cholera outbreak, and received the Chevalier of the Legion of Honor for his contributions. He returned to Paris, and although he had hoped to obtain a post at the Hôtel Dieu, he was not offered one, surely a disappointment to him. But in 1838, the director of a hospital for the deaf (the Institution de Sourds-Muets) died, and that position was given to Menière. Thus, at the age of 39, despite his limited background in otology, he became the head of the hospital, little knowing that this appointment would ultimately lead to his fame (although he was not famous in his life time.) That same year, he married Pauline Becquerel, the daughter of a well-known physicist, Antione Becquerel. About a year after their wedding, their only child, a son named Émile was born to them, who later followed in his father's footsteps and became an otologist.

Menière moved in literary circles, and was friendly with Victor Hugo and Honoré de Balzac. Through such connections, he was acquainted with the Emperor Napoleon III and the Empress Eugénie. He was a person of wide interests that included poetry, history, philosophy, and botany, in addition to medicine. He was a connoisseur of orchids. He was a prolific writer, publishing articles with titles such as "Medical Studies of Some Poets, Ancient and Modern," "Medical Studies of Latin Poets," along with articles on orchids and botany, in addition to numerous publications in the field of medicine.

Since he had not been a specialist in diseases of the ear prior to his appointment, Menière conscientiously set to work learning what was known at the time about its structure and function, and the accepted therapies. One recommended treatment for deafness was catheterization of the Eustachian tubes. Menière himself came to think of deafness as untreatable and considered it best to offer education, such as sign language or lip reading. A surprising range of treatments was employed for patients with vertigo, that sound random and unscientific to us. Many were taken orally, such as quinine, tonics, and soda water. Others were applied to the body such as plasters, oils, and ointments. In some cases, leeches were placed around the ears, or the patient was bled.

In observing his patients, Menière began to notice a correlation between episodic vertigo and deafness in some. He was familiar with the work of Flourens, and thus began to wonder whether the sensation of vertigo arose in the semicircular canals, rather than the brain. In 1848, he observed a young woman who had died shortly after arriving at the hospital, having been admitted after a sudden onset of deafness accompanied by vertigo. On autopsy, he could see blood had invaded the semicircular canals, which gave support to his idea. After familiarizing himself with the histories of several patients, he became convinced that vertigo was a condition that arose in the inner ear. Damage to hearing was associated with it. He also noted that patients often complained of tinnitus, as well. On September of 1861, he presented a paper with his findings to the Imperial Academy of Medicine. It was entitled, "On a particular form of hearing loss resulting from lesions of the inner ear," that included references to Flourens having shown the inner ear to be the seat of balance. The paper was not well received, presumably because it didn't reflect the prevailing understanding of the condition. Unfortunately, Menière died not long afterward in January of 1862, of pneumonia. He was 62.

It was left to his son, Émile Menière, who also became an otologist, to publicize his father's work, and in 1880 he presented a paper at a professional meeting which again described the syndrome, giving it the name Menière. It should be noted that Menière's contribution was to demonstrate that vertigo could originate in the inner ear, and that a connection existed between episodic vertigo and damage to hearing. He did not identify it as a disease entity in the way it is currently understood. Over the course of the ensuing 100 years his understanding became widely recognized and, as is all too often the case, he never had the satisfaction of knowing this.

During the rest of the 19th century, and well into the 20th, more knowledge about vertigo, how it manifests itself, and the conditions that give rise to it developed slowly. Professor Ernst Julius Richard Ewald (1855-1921), a physiologist at the University of Strasbourg, studied the inner ear, and enlarged the understanding of the effects of moving endolymph on movements of the body, head, and eye, and its relation to nystagmus. Like Flourens, he used pigeons for his studies. He was awarded by the Paris Academy of Science for his work in 1892. Robert Bárány (1876-1936) of Vienna was another pioneer in the field, and received a Nobel Prize in 1914 for his contributions in clarifying the physiology of the inner ear, and pathology associated with it. In 1921, he described short-lived episodic vertigo precipitated by head movement, possibly the first person to recognize what would later be seen as a form of BPPV. Both Ewald and Bárány developed tests related to vertigo, some of which still carry their names.

In 1938, Kyoshiro Yamakawa (1892-1980) of Japan was the first person to describe endolymphatic hydrops. About six months later, Charles Hallpike (1900-1979) and Sir Hugh Cairns (1896-1952) of Britain published a similar report. All three had done post mortem studies of the petrous bone, the part of the temporal bone where the vestibular system nestles. In each case, the sample was small, but their work furthered general understanding of the physiology of MD. Interestingly, Yamakawa suffered from MD himself, self-diagnosed in his thirties, before his work was published.

Charles Hallpike, along with Margaret Dix, enlarged the description of BPPV and are credited with giving the syndrome its name. Their work, based on a population of 100 people, was published in 1952. It was at that time that they also described the maneuver they had developed for making the diagnosis, the Dix-Hallpike, discussed in Chapter Three.

During the 1950s and 1960s, research continued. In the 1950s, the invention of the electron microscope made it possible to observe the cilia within the inner ear, and thus gain an enlarged understanding of its functioning. In 1962, (and thus ten years after Dix and Hallpike published their findings) Howard Schuknecht (1917-1996) of Harvard University, using cats as subjects, proposed the idea that detached otoconia (or canaliths) caused the vertigo in BPPV. He had observed material lodged in the posterior canal.

Trolling through the web makes it clear that a significant amount of research, experimentation, and speculation concerning vertigo took place in the middle of the 20th century. The possibility that injury, illness, dental work might cause the conditions were all considered. Many thought that allergic or autoimmune reactions might play a role. By the mid 60s, the distinction between BPPV and MD appears to have been established, as well as the notion that the symptoms of BPPV were attributable to displaced otoconia, while for MD it was the increased pressure caused by endolymphatic hydrops.

Treatment for vertigo limped along. In the 1950s, the use of steroids was considered. Surgery to decompress the endolymphatic sac, or to insert a shunt, began to be performed in that era. Labyrinthectomies (removal of the entire inner ear) were done. Diuretics were recommended as early as the late 30s, based on the understanding that the balance of fluid in the inner ear was the culprit in MD, and the recommendation that a low salt diet be observed seemed to be commonplace at least by the 1970s. Lamentably, as we all know, nothing came up to treat or cure MD that rivaled the success of the Epley Maneuver for BPPV, nor has it to this day.

The development and acceptance of the Epley Maneuver has an interesting history. John Epley, MD, is an otolaryngologist in Oregon, possessed of an inventive mind. Sometime after his residency at Stanford University, he opened a private practice in 1965. Knowing of the research that indicated out-of-place particles were responsible for vertigo in BPPV, he and an assistant, Dominic Hughes (an audiologist,) made a model of the inner ear, using plastic tubes. They inserted BBs in the tubes to serve as the otoconia. They then tried a series of moves, turning their model every which way, to figure out a way to get the particles to go back where they belonged. After settling on a series of motions, Dr. Epley tried using them on some patients, with very good initial results. He continued on, until he'd worked with a large sample, and could feel certain of the effectiveness of the maneuver. In 1980, he first presented his findings at a professional meting in California, along with a demonstration of the technique. His presentation was poorly received. People in the audience walked out, and one man wrote on a comment card, "I resent having to waste my time listening to some guy's pet theory." Three years later, Epley submitted an article about his work to the Journal of Otology, but the paper was rejected. He was unable to find a publisher until 1992, when a report that described 30 cases with a high success rate was placed in The American Academy of Otolaryngology. Even this did not impress everyone, and when he persevered in giving talks to explain the maneuver, audiences were apt to be skeptical or hostile.

Finally, in 1999, The New England Journal of Medicine published a review article by Joseph Furman and Stephen Cass with the simple title, "Benign Paroxysmal Positional Vertigo," describing Epley's Maneuver as the recommended treatment for the condition. By the time that article appeared, others had done clinical trials of the technique, confirming Epley's findings. Thus, nearly twenty years after Dr. Epley first presented his maneuver, it was recognized for its value in a journal read by nearly every physician in the country. Surely a long and distressing twenty years, but at least he saw success for his work in his lifetime.[96] As far as I know, he is still alive, perhaps taking pleasure in the fact that you can scarcely talk about vertigo without saying his name – and certainly taking pleasure in knowing that, thanks to him, countless people no longer live lives plagued by vertigo.

It can be seen that, despite careful observation and studies by both Flourens and Menière, an acceptance by the medical community that a malfunctioning in the inner ear could be the cause of vertigo did not come immediately. Once that idea was finally accepted, it sounds as if research went along in a fairly straightforward way for many years – about 100. Not much hue and cry was raised when physicians tried surgery or standard medication to see what could control or stop vertigo. These were part of the usual medical approaches to any illness. It's not hard to see why suggesting that twisting and turning a person to treat a disease, as in the Epley Maneuver, could seem bizarre – a crackpot's or medicine man's approach. People tend to stick with what is familiar, and to do things the way they are used to doing them. But considering the fact that Epley had good data to support his work, it is a shame that more of his colleagues didn't have an open mind.

Thus, delay in embracing ideas that are not part of the common wisdom in a given era is part of the history of many scientific advances, even for researchers who may be looking for something new. A full understanding usually comes in bits and pieces down over time. People build on the discoveries made in the past, make guesses, draw conclusions that make sense but may be only part of the picture. And

---

96   Most of the information in this paragraph is taken from an article in *The Oregonian* by Joe Rojas-Burke, "Cursing the Cure," which appeared in the 12/31/2006 issue.

once something is found that seems true, it has to be tested on large populations to verify it. I often think of the enormous sample that was needed to prove a strong causal link between cigarette smoking and lung cancer. Or, despite another large sample, the flip-flop that occurred concerning the safety of prescribing replacement hormones for post-menopausal women. Vertigo is not a big enough public health issue, nor a fatal enough one to command the kind of research money that such studies do, which have cast light on cancer or heart disease.

Many of those with whom I spoke (and me, too) feel resentful about what seems to be the lack of interest within the medical world about solving the vertigo riddle. Yet when you plow through reams of articles and Internet sites, you find a confusing and often contradictory array of claims and findings. As is true in almost every area of research, studies conducted by reputable investigators do not necessarily produce the same results. In searching the Web, articles by professionals may recommend the use of (for example) steroids, anti-allergy medication, or betahistine, while others ignore or disparage them. Outside the scientific and professional community, there are numerous postings on the Internet by individuals who give convincing sounding testimony to being helped by a certain substance, diet, or procedure.

As they have done since time immemorial, people are prone to believing in mysterious causes for mysterious phenomena. And vertigo remains a mysterious condition to this day – less so than in Menière's day, but far from clear-cut. It is confusing for those who suffer from it, as it is also for those who try to diagnose and treat it. The history of scientific discovery is replete with accidental findings, as well as findings obtained by carefully constructed research. It is impeded by people's difficulty in seeing things that operate outside the framework of their own understanding. People who raise questions and find alternative ways to conceptualize the workings of an illness are usually the ones who advance research. Research into diseases that are less in the public consciousness needs a special quality of enthusiasm, and unfortunately, most research needs generous funding. Vestibular illnesses need much more of both.

Summary and Afterword

Menière's Disease and Benign Positional Vertigo are experienced as closely related conditions, sharing the symptom of vertigo that appears and disappears in a random and mysterious fashion. For both, the degree of impairment and suffering they inflict varies, although it is safe to say that, with the exception of those whose attacks are very brief and mild, they lurk in the shadows of the mind and often have a significant effect on the course of life. Everyone has to come to terms with it in some way, but when attacks are frequent, severe, and disabling, then that person is faced with adapting to a chronic illness, something more likely to be the case in MD than in BPPV, VN, or labyrinthitis.

Thanks to the wonders of the canalith repositioning maneuvers, in particular the Epley Maneuver, many of those with BPPV are able to live virtually vertigo free lives, or experience it very infrequently. All of those for whom the maneuver works have the comfort of knowing that a future attack can be brought under control. There is no such universal, or nearly universal, treatment for MD, so that each person, hopefully working in concert with a medical practitioner or other professional, will need to learn what works best in his own case. Generally, the mechanisms underlying BPPV are better understood than they are in the case of MD, although what gets an attack of BPPV started is not always so clear. A considerable body of research concerning MD has accumulated over the course of the past 100 years, and it seems to be well established that endolymphatic hydrops is involved in MD, but what makes it come and go, and what to do about it, are far less certain. Even the significance of the hydrops is not crystal clear. In looking through articles written by researchers and physicians about MD, you will find conflicting theories and findings, both as to causes and treatment.

In interviewing people for this book, I was struck with the remarkable variety of symptoms, and combinations of symptoms, that they described. Apparently, their symptoms were confusing not only to them, but in too many cases, to the doctors with whom they consulted as well. It made me wonder whether there are doctors scattered around the country who have little or no familiarity with vestibular illnesses. As many doctors have said, even when a diagnosis is made and a treatment plan instituted, it remains unclear whether or not the treatment is making a difference, since the symptoms eventually resolve on their own. However that may be, the fact there is no surefire and widely accepted treatment for MD creates a difficult situation for doctor and patient alike. For the patient, it is the despair and anger so frequently articulated, the feeling that life is out of control and whatever the doctor suggests isn't really changing much. Doctors, of course, want to be of help and must find it frustrating not to have more to offer.

No matter what theories you subscribe to concerning the underlying causes of these conditions, what makes the symptoms start and stop and what keeps them returning again and again continue to be poorly understood. This is especially true of MD. Because of the intermittent nature of the symptoms, it seems unlikely that endolymphatic hydrops is attributable solely to a mechanical problem such as a narrowing of a shunt. If that were the case, surgery ought to be the definitive treatment. So the question of how and why it develops remains unanswered. As noted in the text, some see a virus as being the culprit. Presumably, greater clarity as to how these diseases work will lead to more definitive treatment. More research, and more closely observed clinical experience, is needed to resolve such questions. Although some research is ongoing, one doesn't get the feeling that large amounts of money are being funneled into it. Some of this is surely because vestibular illnesses are not fatal, and partly because the one likely to cause the most trouble, MD, is relatively rare.

In the future, the subjective experiences of vertigo, tinnitus, and disequilibrium described by patients in varying ways may prove to be indicative of different disease processes that call for different treatments. The spinning room, the churning brains, the vertical scrolling, the bouncing world, the room sliding from one side to the other, the sensation of a precipitous drop - all now fall under the heading of "vertigo." The flapping wings, the scraping metal, the loud roar, the insect buzzing, the high thin note, the distant whisperings are "tinnitus." Difficulty walking a straight line, failing to round a corner, the sense of walking on a rubbery surface, of being pulled to one side are all defined as imbalance, and seem particularly hard to understand when they occur without the accompaniment of vertigo. Right now, they all sit snugly within their headings, the variations not seen as significant, and hence not often inquired about by physicians.

I started this book on a personal note, and will end it on one. Just before the end of the three week course of antiviral medication that began with guarded optimism early in January, 2013, a two week bout of low level symptoms started up, consisting mainly of fatigue, vague nausea, woozy feelings exacerbated by walking around, and double vision upon awakening. Thus it began right when I was still taking the medication! The rest of the winter was clear, with the exception of a couple of days of mild dizziness. April ushered in spring, and also ushered in another ten-day bout of symptoms, with only a four day break before the arrival of yet another bout of about ten days, then another respite, then a mild episode of only three days. Thus, I had about a month wherein I felt crummy more of the time than I felt well. I am hoping for another good long break over the summer when I can half forget about the whole thing. Be that as it may, it is clear that the acyclovir I took had no discernible impact on my condition. I have since learned that antiviral Valtrex (a brand name) is stronger, and might be more effective, but I am not sure I will ask to try it without having received any indication that an antiviral might work for me. After all, I can "live with" my level of symptomatology: low-grade symptoms for a week or two several times a year, often occurring in a cluster. They don't really prevent me from doing much, though I wouldn't ski during one, and find hiking too nauseating. But it is rarely disabling, even though I would of course prefer not to deal with it at all.

Deafness, one of the scourges of MD, has not been a big issue for me. So far as I can tell, my ear took a hit with my first attack, and may have fluctuated some in the early months. But since my "attacks" have switched to being characterized by double vision rather than full-bore vertigo, my hearing has seemed stable to me. Slightly impaired, but no real handicap. The fact that diplopia is thought of as arising in the brain rather than the inner ear may mean that I don't actually have MD – or that I did, but no longer do. Perhaps that could be a good thing and point to a different outcome. I've wondered about migraine-associated vertigo since migraines arise in the brain, but can't see that I have any other symptoms that might be indicative of that diagnosis. In the list of possible diseases characterized by vestibular malfunctioning, I haven't found any that exactly reflect my own experience. But I am not alone in that, as became evident during the time I was interviewing people for this book.

The question of hope arises. The fact that the intensity of my episodes has faded seems a hopeful sign. Each time I've gone for several months with no problem, I begin to believe that it has all finally gone. If I ever went for a couple of years without any symptoms, I would surely feel confident that it had disappeared, despite knowing people can go for years between attacks. Still – hope rises time and time again. Emily Dickinson wrote, "Hope is a thing with feathers/ That perches in the soul/ And sings the tune without the words/ And never stops – at all." The poem's refrain sticks in my head, and speaks to the eternal presence of hope that carries us along through life.

Note on the Spelling of Menière's Name

Very often, you will see Menière written with two accent marks, as follows: Ménière. Occasionally you may see it with no accents, or with an accent only on the first 'e.' Throughout this book you may have noticed I have used an accent mark only over the second 'e.' I did this because of something interesting I came across in my investigations. I found a paper called "Prosper Menière From the Foreign Point of View," authored by O. Michel, included in a collection of articles edited by Olivier Sterkers in the book <u>Menière's Disease Update</u>. The papers in the book were presented at the Fourth International Symposium on Menière's Disease in 1999. Near the end of the article in question, he notes there had been some question about the spelling of Menière's name over the past 100 years. In his research, O. Michel found several letters and other written work signed by Menière himself. All had only the second accent mark, thus showing "clearly and unambiguously" that this is how the man himself spelled his name.

Acknowledgements

I am deeply indebted to Dr. Victor Calcaterra, an otolaryngologist, for reviewing the sections of this manuscript that are devoted to the anatomy and physiology of the ear, as well as the sections describing the features of vestibular diseases and standard treatments for them. He was very generous with his time in talking and communicating with me in order to clarify questions about inner ear anatomy and the characteristics of these illnesses. I thank Dr. David Astrachan, also an otolaryngologist, for meeting with me to answer some questions about vestibular illnesses. Richard Purdy was kind enough to spend time answering questions concerning vestibular physical therapy. Dr. Richard Gacek gave of his time over the phone to talk with me about the possibility that Menière's Disease could be a viral illness. Dr. Jolene Ross was more than generous with her time in talking to me, again over the phone, about biofeedback. My thanks to Cynthia Allen for information about the Feldenkrais Method.

I very much appreciate the assistance with editing provided by Elisa Zonana, John Case, and Diana Perron, who all gave freely of their time, ideas, and perspectives. Madelon Baranoski not only pitched in on the editing, but in addition was extremely helpful to me at the very beginning - before a word was written - in helping me figure out how to organize my material. In addition, she went over the formatting of the bibliography with me. They were all an amazing help.

My sister Marian Willmott designed the fantastic cover for the book. In thinking of how to represent vertigo, it was her idea to use a photograph of trees and distort it. At first I had thought in terms of just using a swirling circle, and her idea was so much more original. She was amazingly able to come up with multiple possibilities for the overall design of the cover. She also drew the diagrams of the ear found in Chapter Five, and was endlessly patient as we tinkered with them. It was fun to work on together on the project, and I can't thank her enough.

My thanks go to Mary-Lou Weisman for encouraging me to think in terms of using a self-publishing company. I had not really considered it before speaking with her.

For their enthusiasm for this project I thank my family: Howard, Elisa, Mark, Jessica, Ariela, Jeremy, Sonja. The same goes for my very good friends: Barbara, Clara, Diana, Elissa, Jo, Natasha, Virginia. Their interest and support has meant more than they may realize.

Many many thanks go out to all the people who agreed to be interviewed for this book and spoke so openly and thoughtfully about their experiences. All of you enlarged my view of vestibular illnesses, as I hope will also be true for those who read this work. Without you, there would have been no book!

Appendix

Recommended Resources

www.dizziness-and-balance.com: Timothy Hain MD seems to be the editor. This is a very comprehensive and well organized website. Dr. Hain, an otolaryngologost. is affiliated with the Chicago Dizziness and Hearing Center, part of Northwestern University. I find this site avoids dogmatism, and provides a thoughtful summary of most topics.

Wikipedia's site, called simply "Menière's Disease," has become increasingly comprehensive. The site for BPPV is much shorter. In searching for any specific sub-topic, it's worthwhile to google it and check out what Wikipedia has to say.

Ed's Meniere's Disease is a series of videos on YouTube offered by an internal medicine physician, Edmund Cheung, who was diagnosed with MD in 2010. Each video covers different aspects of the disease, and includes his personal experiences. You can search for an area of particular interest, or start with video #1 and watch them all. Google "Ed's Meniere's Disease" and a number of them will pop up, not in order. There are blogs under the videos where he has responded to questions people have posted.

Vestibular Disorders Association (VEDA), www.vestibular.org, seeks to help people understand vestibular disorders (not limited to MD and BPPV), live more comfortably with their condition, and offers help, including an on-line support group. You can become a member and receive a newsletter that will keep you informed about issues relating to vertigo. As with MDIC, below, you can find postings of researchers looking for subjects for clinical trials.

Menière's Disease Information Center (MDIC), www.menieresinfo. com, is a site maintained by non-professionals as a public service. It has an organized list of areas of interest, such as symptoms, diagnosis, treatment, etc. It offers links to articles, often very recent ones, such as research in progress. It also posts announcements of researchers looking for subjects for clinical trials.

Menière's Disease: What You Need to Know, by P.J. Haybach, RN, MS is a comprehensive book written for the layman. Originally published in 1998, it remains a valuable resource, describing in detail symptoms, tests, how things work. Written by a nurse, it is far more medically detailed than this book. Obviously, the reader will be aware that it doesn't include information and treatments that have emerged since it was published. It is available through VEDA or Amazon. Ms. Haybach also wrote a book on BPPV, published in 2000, but this is not easy to find.

Hints for Dealing with Challenges
of Vertigo and Imbalance

The following are hints for dealing with some of the many issues that come up when you live with vestibular illnesses. Some are from my own experience, others were offered by the people I interviewed. Some are probably just common sense, but you might find something helpful in here.

<u>Climbing the stairs</u>: Before I had vertigo, I never realized that you automatically lean forward when you go upstairs. I found I had to make a conscious effort to do this, to avoid stepping back into thin air as I tried to catch my balance - an alarming thought. And of course use the railing on the stairs, either going up or down.

<u>Carrying packages</u>: Carrying a weight throws you off balance, more than I would have imagined. It's good to be mindful of this, especially on the stairs. On flat ground, even out your burden if you can – one half for each hand.

<u>Walking straight #1</u>: When you tend to stagger around like a drunken sailor, it helps to fix your eyes on a distant point and walk toward it, holding your head steady.

Walking straight #2: Any time you find yourself stumbling around, it helps to stop moving, square your shoulders, and then proceed on with your eyes fixed on the distance, as above. If you're so unsteady you're about to keel over, none of this will make much difference, however.

Driving: If you experience a wave of dizziness while driving, again fix your eyes on something well ahead of you. This will make it easier to pull the car to the side of the road without disaster. Early on, a doctor suggested this to me, and I found it reassuring to bear in mind.

Sleeping: When woozy or subject to peculiar feelings in your head, it helps to sleep with the affected ear *up*. The day after I read about this, I was in bed lying on my stomach with the "good" ear up. Along the top of my head, I had a sensation of waves rolling onto the shore, over and over. I didn't think I could fall asleep with that going on, and was afraid it would lead to throwing up. I turned my head in the other direction, with the "bad" ear up. As if by the flick of a switch, the sensation stopped immediately.

Avoiding triggers: Some people find that certain triggers can set off an attack. Noticing what these are, and then - of course - trying hard to avoid them makes good sense. Common triggers are tipping your head back, turning over in bed, sudden changes of position in general, loud noises, strong scents in the supermarket, things moving or rushing near or around you, a lot of sensory stimulation, confusion. When in a situation where you need to change position in a way that's likely to make you dizzy, it helps to move *slowly*.

Loud noises: Some people keep earplugs with them, to use in situations where they might be overwhelmed by noise, due to hyperacusis. Two people found they were able to enjoy a concert or a band playing if they used them.

Attitude: Many people spoke of the importance of trying to forge ahead and do whatever you are able to do, trying to prevent the condition from dominating your life. Try to immerse yourself in hobbies, interests, and activities when you are able. Some people needed to change jobs, either because of issues directly related to vertigo, or because of hearing problems, and were able to do this in an adaptive way. One person said, "I refuse to let something like this affect my life." Another used the "spoons" idea, from a book by Christine Miserandino, The Spoon Theory, which addresses dealing with a chronic illness. In a nutshell, the book states that you know the possibilities for your day may be limited by your condition. You may not have enough energy to do all that you would like. So you plan in the morning how you want to focus what energy you do have. In this way you feel in better control, more in charge of your life. Many people emphasized how important it is to take charge of your own care, to be informed, to question things, to investigate possibilities. "You have to be your own advocate" was said to me more than once.

Communication: One of the trials of vertigo is the effect it can have on you emotionally. Although no one wants to complain repeatedly about their symptoms, many people said it helped when there was someone in their lives they could talk to about it. In trying to speak with a person close to you, it may help to let that person know what kind of response feels helpful. For example, if you only need someone to listen and aren't looking for advice, you could explain that. It also felt helpful to find others who had problems with it. Some people liked being part of a support group – usually this was online. Some can be found through the Vestibular Disorders Association (VEDA.)

Anxiety: This is common, both during vertigo episodes, and between them as well. Learning to manage anxiety is important. Some people learned to practice deep breathing, being "in the moment" (although not in the middle of an attack!), meditation, yoga. Vestibular physical therapy as well as tai chi and taekwondo help, and are helpful for balance as well. Some found the use of medication helpful in controlling anxiety.

Information: Many people – myself included – felt better after learning more about vertigo. Various books and web sites are listed in the "Resources" pages here in the appendix. The bibliography has many references, but in a less accessible form. "Googling" the ones of interest to you may sometimes work out better than trying to access the topic through the listed website.

Glossary and Lingo

Abducens nerve – The 6th cranial nerve, running from the brainstem to the eye, part of the vestibulo-ocular reflex (VOR). This is one of the motor nerves that control movements of the eye.

Acoustic nerve – This is an older term for the nerve leading from the inner ear to the brainstem, now called the vestibulocochlear nerve. It carries the signals related to balance from the semi-circular canals as well as the auditory signals from the cochlea. Eighth cranial nerve.

Acyclovir – An anti-viral drug, which some believe to be effective in treating MD and vestibular neuritis. Brand name: Zovirax.

Adrenaline – Hormone released by the adrenal glands when the individual senses danger. Along with cortisol and other hormones, produces the "fight-or-flight" reaction.

Alacyclovir – Another anti-viral drug, which may be more effective against viruses than acyclovir. Brand name: Valtrex.

Antiemetic – Medication used to control nausea and vomiting.

Antivert – Brand name of a drug to treat motion sickness.

Anvil – One of the three bones of the middle ear. Scientific name: incus.

Auditory nerve – An earlier name for the vestibulocochlear nerve. (See "acoustic nerve" above.)

Autonomic nervous system – The part of the nervous system that operates automatically, with little or no conscious control, thus including breathing, heartbeat, digestive tract, etc.

Benzodiazapine – A class of drugs used to calm anxiety, such as Xanax, Valium.

Betahistine – Developed in Europe in 1970 to treat MD, it dilates the blood vessels in the inner ear.

Bilateral – Means anything that takes place on both sides of the body

Bonine – Brand name for a medication for dizziness or for motion sickness.

Brain stem – The part of the brain at the very back of the head, perched on top of the spinal cord. It regulates most of the automatic functions of the body.

Burn out – A phrase used to denote a time when Menière's Disease has destroyed the cilia in the canals, so the individual is no longer subject to vertigo attacks. It would occur after years, is not universal.

Canaliths – Tiny calcium crystals found in the utricle and saccule, known as otoconia when in place. Called "canaliths" when they migrate to the semicircular canals.

Canalith Repositioning Maneuver – Maneuver intended to put displaced canaliths back where they belong, similar to the Epley Maneuver.

Canalithiasis – Condition wherein the canaliths are floating in the semi-circular canals where they don't belong, as in BPPV.

Cilia – Miniscule hairs. In the inner ear, large numbers line the canals in the cochlea and the semicircular canals, where they transmit signals to the brain via the vestibulocochlear nerve.

CNS – central nervous system.

Cochlea – The part of the inner ear that receives and transmits auditory signals. It is shaped a bit like a snail or conch shell.

Compazine – Medication sometimes used to control nausea, although its original use was as an antipsychotic. It also helps to reduce anxiety.

Cortisol – One of the "stress" hormones, part of the endocrine system, released by the adrenal glands atop the kidney.

Cranial nerves – Nerves that arise in the brain, not the spinal cord. There are 12 of them.

Cupula –The sense organ at the base of the semicircular canals that transmits information about balance to the vestibular nerve.

Cupulolithiasis – Condition wherein otoconia have moved out of the otolith organs and get lodged in the cupula, usually that of the posterior canal, causing positional vertigo.

Dexamethasone – A steroid sometimes used to treat vertigo.

Diaphoresis – This is the technical term for sweating!

Diplopia – The technical term for double vision.

Dix-Hallpike Maneuver – Similar to the Epley Maneuver, but less elaborate. Used to determine whether an individual might have BPPV, or to verify how effective an administration of the Epley was.

Diuretic – A medication that increases the production of urine. Often called a "water pill."

Drop attack – An attack that strikes so suddenly that the individual instantly falls to the ground. Usually only occurs in later stages of vestibular illness.

Endolymph – The fluid found within the semicircular canals.

Endolymphatic hydrops – An excess of endolymph.

ELH – Abbreviation for endolymphatic hydrops.

ENT – An ear, nose, and throat physician, also known as an otolaryngologist.

Eustachian tubes – These connect the middle ear with the back of the nose. They are usually closed, except when you swallow or yawn. They keep the pressure equal on both sides of the eardrum.

Epley maneuver – A specialized maneuver to get displaced canaliths back where they belong, or where they won't cause vertigo.

Fukada stepping test – A test to check a person's equilibrium. The patient crosses arms over chest, closes eyes, and marches in place. If individual is very affected, will have trouble maintaining balance, if less affected will march in a small circle. If all is well will maintain original position.

Ganglion – A collection of nerve cell bodies. The nucleus of the cell, among other things, is found within the nerve cell body. The vestibular ganglion is just outside the vestibule.

Gans Maneuver – Another canalith repositioning maneuver. Not in much use in the US.

Gentamicin – An antibiotic that has the curious side effect of destroying the cilia in the inner ear. It is administered via transtympanic injection (through the eardrum). Because it permanently destroys the functioning of the semicircular canals, it is only used in cases where the individual is gravely affected by vertigo. The risk is that it may destroy the hearing as well. Several small doses have been found to reduce that risk.

Hammer – One of the three bones of the middle ear. It articulates with the eardrum. Scientific name: malleus.

Hydrochlorothiazide – A diuretic, commonly used to treat MD. Originally developed as a medicine for high blood pressure.

Hyperacusis – Hypersensitivity to particular loud sounds – a paradoxical reaction that may accompany hearing loss. See "recruitment."

Idiopathic – An adjective meaning the cause of a disease or medical problem is unknown.

Incus – Technical term for one of the three bones in the middle ear, called the "anvil." It's the one in the middle.

Labyrinth – A word for the inner ear, so called because it resembles a maze.

Labyrinthectomy – Surgical removal of the entire inner ear.

Laryngologist – An ear, nose, and throat physician (ENT) who specializes in the throat.

Lempert Maneuver – A canalith repositioning maneuver for use when the canaliths are situated in the horizontal semicircular canal. Also called the "barbecue roll."

Malleus – Technical term for one of the three bones in the middle ear. Colloquial name: "hammer." It connects to the eardrum.

Mastoid – Posterior part of the temporal bone.

MAV – Migraine Associated Vertigo. Recent research indicates some vertigo may be associated with migraine, possibly even when the patient does not have headaches.

Meclizine – Medication used to treat motion sickness, very commonly recommended as a treatment for vertigo.

Meniett Device – An instrument designed to treat vertigo by changing the pressure in the inner ear. It requires the insertion of a port (opening) through the eardrum. Held in higher regard in Europe than in the US.

Neuronitis – Vestibular neuritis (VN.) A viral infection in the vestibular nerve and/or the vestibular ganglion.

Neurotransmitter – A chemical substance that transmits nerve impulses across a synapse.

Neurotologist – An otolaryngologist who has specialized training in the neurological conditions of the ear.

Nystagmus – Involuntary rapid eye movements that accompany vertigo.

Oculomotor nerve – The 3rd cranial nerve. It's a motor nerve that runs from the brainstem to the muscles of the eyeballs. Part of the vestibulo-ocular reflex (VOR).

Oscillopsia – The sensation that stationary objects are moving around, or bouncing.

Ossicles – The three tiny bones in the middle ear that carry acoustic impulses to the cochlea: the malleus, incus, and stapes.

Otoconia – Tiny crystals within the utricle and saccule, in the otolith organ.

Otolaryngologist – An ear, nose, and throat physician, or ENT.

Otolith organs – The organs situated in the vestibule at the base of the semicircular canals, where the utricle and saccule are found. It is where the otoconia reside. They monitor the body's relationship to gravity and linear movement.

Otologist – A subspecialty within the ENT field. An otolaryngologist who specializes in the ear.

Oval window – Membrane covering one end of the cochlea, through which sound vibrations are transmitted. The stapes bone adjoins it.

Perilymph – Fluid found in the canals within the cochlea, different in chemical composition from the endolymph.

Petrous bone – Connected to the mastoid bone, and part of the temporal bone. It's where the inner ear is situated.

Pinhole leaks – Ruptures, presumably due to heightened pressures, that may occur in the membranes lining the semicircular canals and cochlea.

Pinna – The external ear. (Plural: pinnae)

Proprioception – Feedback system within the body, wherein the position of muscles and the body's position in space are monitored. Operates unconsciously.

Recruitment – When some auditory cilia are damaged, other cilia may be "recruited" to respond to sounds, causing them to sound louder than they should. May produce a sound that is actually painful (hyperacusis.)

Round window – Membrane covering the far end of the coiled canal in the cochlea. It bulges in and out in response to pressures within.

Saccades, saccadic movement – Quick, jerky movements of the eye, as in nystagmus.

Saccule – One of the two sacs situated at the base of the semi-circular canals that contain the otoconia. The other is the utricle.

Scopolamine patch – Drug used to prevent or address motion sickness, may be worn on the skin as a patch.

Selective serotonin reuptake inhibitors – Known as SSRIs, these are the most commonly used anti-depressants today. Serotonin is a neurotransmitter that improves mood. These drugs block the reuptake of the serotonin so that it remains active longer in the synaptic space.

Semicircular canals – The three looped canals that arise from the vestibule, next to the cochlea, that regulate balance.

Semont Maneuver – A canalith repositioning maneuver.

SSRI – See "Selective serotonin reuptake inhibitors."

Stapes – Bone in the middle ear that articulates with the oval window to carry auditory information to the cochlea. The technical term for "stirrup."

Stirrup – Colloquial term for the "stapes," above.

Synapse – Space between two nerve endings. Neurotransmitters carry nerve impulses across them.

Syndrome – Group of symptoms that occur together that are characteristic of a certain illness or condition.

Temporal Bone – The bone in which the middle and inner ear are located.

Tinnitus – Sound generated within the ears. Often a steady high note, but may be rhythmic, metallic, vary in tone. Usually accompanies MD.

Transtympanic – Means "through the eardrum," and in this book refers to medication that may be administered in that way.

Trochlear nerve – One of the three cranial nerves that run from the brainstem to the muscles of the eyes. It is cranial nerve #4.

Tricyclic antidepressants – These are the older type of antidepressant, in less use since the development of the SSRIs. They may be helpful in treating MAV.

Tumarkin crises – *or falls* – A fall that occurs without the sensation of vertigo, so suddenly that a standing person will flop to the ground. If unhurt by the fall, will recover equilibrium right away, without aftereffects.

Tympanum – The middle ear.

Tympanic membrane – Technical term for eardrum.

Typanostomy – Surgical procedure for making an opening in eardrum.

Utricle – One of the two sacs in the vestibule that contain the otoconia. The other is the saccule.

Valtrex – Brand name of an antiviral drug, valacyclovir.

Vestibular – Of or pertaining to the part of the inner ear that regulates balance.

Vestibular nerve –The vestibulocochlear nerve.

Vestibular neuritis – Inflammation or infection of the vestibular nerve and/or the vestibular ganglion. Also known as neuronitis.

Vestibular ganglion – The sensory ganglion of the vestibular part of the eighth cranial nerve. "Sensory" means that it sends information from the senses to the brain. The ganglion is located just outside the vestibule.

Vestibulo-ocular reflex (VOR) – The interaction between the nerve impulses sent to the brainstem via the vestibulocochlear nerve and out to the eyes via cranial nerves #3and #6.

Vestibule – The space in the bone that contains the vestibular organs.

Vestibulocochlear nerve – The nerve that carries both auditory and vestibular information to the brain from the inner ear. The 8th cranial nerve, a sensory nerve.

Vestibular Physical Therapy Q&A

The following are questions I asked of Rick Purdy, a vestibular physical therapist, along with his answers.

Q: Would you describe the ways in which Vestibular Rehabilitation can help people who suffer with vertigo?

A: Vestibular rehabilitation helps in the following ways:

1. Treats BPPV and teaches how to manage the condition in case of a recurrence. (There is no evidence suggesting that BPPV can be prevented through prophylactic measures.)
2. Decreases or eliminates dizziness from conditions other than BPPV.
3. Improves visual stability with head movement. This is the primary function of the vestibular system.
4. Decreases motion sensitivity.
5. Improves balance and postural stability and therefore safety.
6. Teaches how to manage other common symptoms such as light or noise sensitivity, or difficulty navigating supermarket aisles and other visually stimulating environments, to name a few.

The goal of vestibular therapy is to restore patients as much as possible to their previous levels of activity as soon as possible. One of the more important aspects of rehabilitation is to educate patients concerning the cause of their symptoms. Many people dealing with deficits of the vestibular system have difficulty explaining how they are feeling and commonly state that their family and friends do not understand.

Q: Do you distinguish which of the three semicircular canals is affected by displaced canaliths? If so, does this change anything about administering the Epley Maneuver?

A: Yes and yes. The PT needs to figure out which canal the displaced canaliths have moved into, or are "stuck" in, so as to determine which treatment should be administered. The Epley Maneuver (as it was originally developed) only treats free-floating canaliths in the posterior canal. Treatment techniques such as the Semont, Liberatory, and Appiani Maneuvers address some of the other forms of BPPV. An interesting fact is that most practitioners administer the Canalith Repositioning Procedure (CRP) rather than the Epley Maneuver.

Q: I've read about "gaze-stabilization" exercises. I've also read about "habituation" as a method for de-sensitizing a person to a position or action that tends to precipitate vertigo. Are these commonly used?

A: Treatment depends on the cause of the symptoms. For example, when a person with BPPV is experiencing vertigo, there is no benefit to working on gaze-stabilization exercises. The movement of the eyes (nystagmus) cannot be volitionally controlled. In cases of central vestibular dysfunction, in contrast to peripheral dysfunction, working on habituation activities may only make the person nauseous and achieve no rehabilitative benefit.

Choosing which techniques to administer requires a detailed evaluation of the cause, but those you mentioned are commonly used.

Gaze-stabilization exercises are usually prescribed when one or both sides of the vestibular system suffer a loss of function that affects the vestibulo-ocular reflex (VOR). The VOR helps keep vision stable while the head is moving.

The use of habituation exercises presupposes an intact brain (central vestibular system/cerebellum) in order for the body to learn to reduce motion sensitivity. This concept requires frequent repetition of motion to desensitize the body's response to position changes.

Q: Am I correct in assuming that balance training is mainly to help people compensate for damage to balance caused by episodes of vertigo?

A: Episodes of vertigo do not necessarily produce balance *damage*. During periods of vertigo balance is temporarily compromised but some people with BPPV don't develop balance problems that *last*. Imbalance can be caused by damage or other change to any of the three main systems responsible for balance: the vestibular, visual, and proprioceptive systems.

In addition, issues affecting the cerebellum affect the ability of the brain to improve balance. Body strength (especially in the legs), neuropathy, and other concomitant factors also play roles. I have even seen severely worn shoes cause imbalance. This needs to be assessed by a Physical Therapist.

Q: Do you work primarily with people diagnosed with BPPV or is some of what you do also applicable to those with other diagnoses such as Menière's?

A: Others too. BPPV is the more common diagnosis but I also frequently see people with vestibular labyrinthitis or neuritis that resulted in a loss of function. I also see people with acoustic neuromas, concussive vestibulopathy/trauma, disuse disequilibrium, peripheral neuropathy, etc. I will work with people with Menière's between episodes when there has been a loss of vestibular function and the symptoms are primarily motion induced.

Q: I've talked to several people who regularly go to PT when they experience an attack of vertigo. Are you in favor of training patients to do the Epley or other canalith repositioning maneuvers at home? Is it sometimes too difficult for people to do these effectively on their own?

A: I frequently teach patients home techniques to treat BPPV but I make sure they can perform them correctly. These techniques are not difficult to perform but there are important details to keep in mind or else the treatment will be ineffective and (worse) possibly even cause an additional form of BPPV. Some patients are afraid to perform maneuvers at home because of the severity of their vertigo so I encourage them to have a family member assist or seek help from a vestibular PT or MD.

Q: Do you find a prophylactic benefit in a person's doing such procedures at home on a routine basis?

A: Research has not shown the benefit of prophylactic performance of these exercises in decreasing the recurrence of symptoms.

Q: Where can you find a PT who specializes in Vestibular Rehab and does insurance cover it? Do some people come directly to you without a referral from a physician?

A: You can find a list of therapists online at VEDA (Vestibular Disorders Association).

Any practitioner can sign up to be listed regardless of training. Emory University in Atlanta has a certification course and a list of practitioners who have passed the examination in Vestibular Rehabilitation. Our office regularly receives random calls from prospective patients inquiring about our training, so you can also check out local practices.

Insurance does cover treatment since a great deal of research has demonstrated the benefit of vestibular rehabilitation. However, most insurance plans in CT require a referral from a physician in order for payment to be made. The amount of coverage varies.

Bibliography

During May, June, and July of 2013, I checked on the web addresses listed below. At that time, they were accessible, although for many of the entries simply googling the reference is an easier way to find it.

Arenbert IK *et al.* "Vincent's Violent Vertigo." <u>Acta Otolaryngolic Supplement</u>. 1991, Vol.485, pp 84-103.

Ataman T & Anache A. "Our Experience With the Medical Treatment of Endolymphatic Hydrops." <u>International Tinnitus Journal</u>. Jul-Dec 2007, Vol.13, No.2, pp 138-142.

Baloh, Robert W. "Prosper Ménière and His Disease." Jul 2001, <u>Archives of Neurology</u>, Vol.58, No.7.

Baloh, Robert W. "Vestibular Neuritis." <u>The New England Journal of Medicine</u>. Mar 2003, Vol.348, pp 1027-1032.

Balaban, Carey D. & Jacob, Rolf. "Background and History of the Interface Between Anxiety and Vertigo." <u>Journal of Anxiety Disorders</u>, 2001, Vol.15, pp 27-51.

Bartels, Loren. "Vestibular Neuritis & Viral Labyrinthitis." <u>Tampa Bay Hearing & Balance Center</u>. <u>www.tampabayhearing.com/vestibularneuritis.php</u>

Berlinger, Norman. "Meniere's Disease: New Concepts, New Treatment." <u>Minnesota Medicine</u>. Nov 2011, Vol.94, No.11, pp 33-36.

Bienfang, Don. "Overview of Diplopia." <u>UpToDate</u>. Literature Review through 12/2012. <u>www.uptodate.com/contents/overview-of-diplopia</u>

Bladin, Peter. "History of 'Epileptic Vertigo': Its Medical, Social, and Forensic Problems." Epilepsia. Apr 1998, Vol.39, No.4, pp 442-447.

Blakley, Brian & Siegel, Mary-Ellen. Feeling Dizzy. Wiley Publishing Inc., 1995.

Blumer, Dietrich. "The Illness of Vincent VanGogh." American Journal of Psychiatry. Apr 2002, Vol.159, No.4, pp 519-526.

Brandt, T et al. "Long Term Course & Relapses of Vestibular & Balance Disorders." Institute of Clinical Neurosciences, Germany. 2010, Vol. 28, No.1, pp 69-82.

Brody, Jane. "Living With a Sound You Can't Turn Off." New York Times. Tue, 12/04/2012.

Brody, Jane. "A Stable Life, Despite Persistent Dizziness." New York Times. Tue, 01/15/2008.

Casani, AP et al. "Intratympanic Treatment of Intractable Unilateral Meniere's Disease: Gentamicin or Dexamethasone?" Otolaryngology Head and Neck Surgery. Mar 2012, Vol.146, No.3.

Cheung, Edmund. "Meniett Device." Ed's Meniere's Disease. YouTube. www.youtube.com/watch?v+hj-vBFcTKcs

Cheung, Edmund. "Ed's Meniere's Disease: Introduction." Ed's Meniere's Disease. YouTube. http://www.youtube.com/watch?v=youQhkZeZuY

Cheung, Edmund. "Long Term Course of Meniere's Disease." Ed's Meniere's Disease. YouTube. http://www.youtube.com/watch?v=_M0xJu2gwiM

Cima, Rilana et al. "Specialised Treatment Based on Cognitive Behavior Therapy versus Usual Care for Tinnitus: a Randomized Controlled Trial." The Lancet. May 2012, Vol. 379 (9830), pp 1951-1959.

Cohen B. "Inner Ear Troubles: The Roar in the Forest?" The FASEB Journal. May 2006, Vol.20, No.7 pp 806-808.

Conlon, B & Gibson, W. "Ménière's Disease: Incidence of Hydrops in Contralateral Asymptomatic Ear." The Laryngoscope. Nov 1999, Vol.109, No.11, pp 1800-1802.

Cowan, Alan, "Treatments for Ménière's Disease." Grand Rounds UTMB, 12/13/2006.

Crane, Randy. Overcoming Meniere's Disease: A Practical Guide." Self Published 2011.

Devaiah, AK & Ator, GA. "Clinical Indicators Useful in Predicting Responses to Medical Management in Meniere's Disease." The Laryngoscope. Nov 2000, Vol.110, No.11, pp 1861-1865.

Dinces, Elizabeth & Rauch, Steven. "Meniere Disease." (Review of literature through March 2013.) UptoDate website. www.uptodate.com/contents/meniere-disease

Disabled World (website.) "Famous People with Meniere's Disease," 2008, www.disabled-world.com/artman/publish/menieres-famous.shtml

Enchanted Learning (website.) "Ear Anatomy." www.enchantedlearning.com/subjects/anatomy/ear/

Fishman, J.M. *et al.* "Corticosteroids for the Treatment of Idiopathic Acute Vestibular Dysfunction (Vestibular Neuritis)." The Cochrane Library (Wiley Online Library), May 2011.

Franz, B *et al.* "P-100 in the Treatment of Meniere's Disease: a Clinical Study." International Tinnitus Journal. Jul-Dec 2005, Vol.11, No.2, pp 146-149.

Gacek, Richard R. "Menière's Disease is a Viral Neuropathy." Journal for Oto-Rhino-Laryngology (ORL). Jan 2009, Vol.71, pp78-86.

Gacek, Richard R & Gacek, Mark. "Menière's Disease as a Manifestation of Vestibular Ganglionitis." American Journal of Otolaryngology. Jul-Aug 2001, vol.22, pp 241-250.

Gates, GA *et al.* "Meniett Clinical Trial: Long Term Followup." Archives of Otolaryngology Head & Neck Surgery. Dec 2006, Vol.132, No.12, pp 1311-1316.

Gates, P. "Hypothesis: Could Meniere's Disease Be a Channelopathy?" Internal Medicine Journal, Aug 2005, Vol.35, No.8, pp 488-489.

Gill-Body, Kathleen. "Current Concepts in the Management of Patients With Vestibular Dysfunction." 2007. (pdf available online.)

Greenberg, SL & Nedzelski, JM, "Medical and Noninvasive Therapy for Menière's Disease." The Otolaryngologic Clinics of North America, Oct 2010, Vol.43, No.5, pp 1081-1090.

Gurkov R, *et al.* "Endolymphatic Hydrops in the Horizontal Semicircular Canal." The Laryngoscope. Feb 2013, Vol.123, No.2, pp 503-506.

Guyot, JP *et al.* "Intratympanic Application of Antiviral Agent for Treatment of Meniere's Disease." Journal for Otorhinolaryngology & Its Related Specialities. Feb 2008, Vol.70, No.1, pp 21-26.

Hain, Timothy C. "Benign Positional Paroxysmal Vertigo." Oct 2012. http://www.dizziness-and-balance.com/disorders/bppv/bppv.html

Hain, Timothy C. "Destructive Treatments of Vertigo." American Hearing Research Foundation. Jul 2007. www.american-hearing.org/disorders/central/destructive_treatment_vertigo.html

Hain, Timothy C. "Dizziness and Hearing Loss - Is It Menière's?" Dizzy Doctor Website. Nov 2010. www.tchain.com/cv/hain-t.htm

Hain, Timothy C. "Meniere's Disease." <u>American Hearing Research Foundation</u>. www.american-hearing.org/name/menieres.html (Page modified 10/12)

Hain, Timothy C. "Migraine Associated Vertigo (MAV.)" Jul 2013. http://www.dizziness-and-balance.com/disorders/central/migraine/mav.html

Hain, Timothy C. "Surgical Treatment of Vertigo." Feb 2011. www.dizziness-and-balance.com/treatment/surg.html

Hain, Timothy C. "Vestibular Neuritis and Labyrinthitis." Sep 2012. www.dizziness-and-balance.com/disorders/unilat/vneurit.html

Hawkins, Joseph E. "Sketches of Otohistory, Part 5: Prosper Ménière: Physician, Botanist, Classicist, Diarist and Historian." <u>Audiology Neuro-Otology</u> 2005; Vol.10. pp 1-5.

Haybach, P.J. <u>Meniere's Disease: What You Need to Know</u>. Published by Vestibular Disorders Assoc, 1998.

Haybach, P.J. "Ménière's Disease." VEDA Publication No. F-4. Undated.

Hill S, Digges N, Silverstein H. "Long-term Follow-up After Gentamicin Application via the Silverstein MicroWick in the Treatment of Meniere's Disease." <u>Ear, Nose, & Throat Journal</u>. Aug 2006, Vol.85, pp 494,496,498.

Hornibrook, Jeremy. "Benign Paroxysal Positional Vertigo: History, Pathophysiology, Office Treatment and Future Directions," (Review article) <u>International Journal of Otolaryngology</u>, May 2011, Vol.2011, Article ID 835671.

Huppert D, Strupp M & Brandt T. "Long Term Course of Meniere's Disease Revisited." <u>Acta Oto-Laryngologica</u>. Jun 2010, Vol.130, No.6, pp 644-651.

John of Ohio. "A New Approach to Meniere's Disease, The John of Ohio Meniere's Regimen." Undated. www.menieresfoundation.org/johnsregimen.htm

Johns Hopkins Medicine Website. "Neurology and Neurosurgery – Vestibular Disorders – Labyrinthitis." No date given. Cited 7/2013. http://www.hopkinsmedicine.org/neurology_neurosurgery/specialty_areas/vestibular/conditions/labyrinthitis.html

Kleinman, Arthur. <u>The Illness Narratives: Suffering, Healing & the Human Condition</u>. Basic Books, 1988.

Lacour, Michel *et al.* "Betahistine in the Treatment of Meniere's Disease." <u>Neuropsychiatric Disease and Treatment</u>. Aug 2007, Vol.3, No.4, pp 429-440.

Lanska, D. "Clinical Summary, Drop Attacks." <u>Medlink (Neurology)</u>, updated May 2012

MayoClinic.com. "Dizziness." May 2006. www.mayoclinic.com/health/dizziness/DS00435/DSECTION=3

MayoClinic.com. "Migraine." Undate. Last cited Jul 2013.

MayoClinic.com. "Stress Management." www.mayoclinic.com/health/tai-chi/

McCall, AA *et al.* "Drug Delivery for Treatment of Inner Ear Disease: Current State of Knowledge." Ear Hear. Apr 2010, Vol.31, No.2, pp 156-165.

MDGuidelines (website.) "Vestibular Neuronitis." No date or author. Date cited 7/13. http://www.mdguidelines.com/vestibular-neuronitis

MedicineNet.com (website). "Hearing & Balance Anatomy." www.medicinenet.com/script/main/art.asp?articlekey=21685

Megerian, CA *et al.* "Surgical Induction of Endolymphatic Hydrops by Obliteration of Endolymphatic Duct." Journal of Visualized Experiments. Jan 2010, Vol.35.

Meniere's Australia (website) www.menieres.org.au/menieres-disease.php

Meniere's Disease Information Center (Website.) "Atypical Meniere's Disease." Undated. Cited Jun 2012. http://www.menieresinfo.com/symptoms.html

Meniere Man. Vertigo Vertigo. Page Addie, 2012.

Merck Manual (website) "Dizziness and Vertigo." www.merckmanuals.com/professional/ear_nose_and_throat_disorders/approach_to_the_patient_with_ear_problems/dizziness_and_vertigo.html (or google "merck manual dizziness and vertigo.")

Millard, Candace. Destiny of the Republic: A Tale of Madness, Medicine, and the Murder of a President. Anchor Books, 2012.

Mira E. "Betahistine in the Treatment of Vertigo." Acta Oto-laryngologica. Jun 2001, Vol.21, pp 1-7.

Morales, A *et al.* "Vestibular Drop Attacks or Tumarkin's Otolithic Crisis in Patients with Meniere's Disease." Acta Otorrinolaringologica Espagnola. Dec 2005, Vol.10, pp 469-471.

National Health Service, Website NHS – Choices. "Labyrinthitis." No date given. http://www.nhs.uk/Conditions/Labyrinthitis/Pages/Treatment.aspx

National Institutes of Mental Health: National Institute on Aging. "Inside the Human Brain." www.nia.nih.gov/alzheimers/publication/part-1-basics-healthy-brain/inside-human-brain

Nobelprize.org (website). "Robert Bárány Biographical." http://www.nobelprize.org/nobel_prizes/medicine/laureates/1914/barany-bio.html

Nuland, Sherwin. "Lia Lee Obituary." New York Times. 9/14/2012

Oboudiyat, Carly. "Dix-Hallpike Positive, No Red Flags, Now What?" Clinical Correlations. (NYU Online Journal of Medicine) Jan 23, 2010.

Odkvist, L. "Pressure Treatment vs Gentamicin for Meniere's Disease." Acta Otolaryngology. Jan 2001, Vol.121, No.2, pp 266-268.

Otolaryngology E News (website). "Prosper Ménière: An Apostle of Humility." Mar 8, 2011. http://www.drtbalu.co.in/news/article.php?story=20110308105233591

Perez-Garrigues, H et al. "Time Course of Episodes of Definitive Vertigo in Meniere's Disease." Archives of Otolaryngology Head & Neck Surgery. Nov 2008, Vol.134, No.11, pp 1149-1154.

Phillips, John & Prinsley, Peter. "A Unified Hypothesis for Vestibular Dysfunction?" Otolaryngology – Head and Neck Surgery. 2009, Vol.140, pp 477-479.

Poe, Dennis (Editor). The Consumer Handbook on Dizziness and Vertigo. Auricle Ink Publishers, 2005.

Pullens, B et al. "Surgery for Meniere's Disease." Cochrane Database System Review. Jan 2010, No.1, CD005395

Rauch, Steven. "Clinical Hints & Precipitating Factors in Patients Suffering from Meniere's Disease." Otolaryngological Clinics of North America. Oct 2010, Vol.43, No.5, pp 1011-1017.

Rauch, Steven. "Dr. Rauch's Online Otology Clinic." Massachusetts Eye & Ear Website. Undated. Cited 7/2013. http://www.masseyeandear.org/about-us/videos-and-podcasts/rauch/gentamicin/

Rojas-Burke, Joe. "Cursing the Cure: Doctor and Invention Outlast Jeers and Threats." The Oregonian. Sun Dec 31, 2006.

Safran, AB et al. "Vestibular Neuritis: a Frequently Unrecognized Cause of Diplopia." Klin Monbl Augenheilkd. May 1995, Vol.206, No.5, pp 413-415.

Salt, Alec. "Cochlear Fluids Research Laboratory." Washington University School of Medicine. http://oto2.wustl.edu/cochlea/

Scholarpedia: Vestibular System http://www.scholarpedia.org/article/Vestibular_system

Schuknecht, Harold. Pathology of the Ear. First published 1974. Second edition.

Shichinohe, Mitsue. "Effectiveness of Acyclovir on Meniere's Syndrome – observation of clinical symptoms in 301 cases." Sapporo Medical Journal, Dec 1999, Vol.68, No.4-6, pp 71-77 (abstract)

Silverstein, H. "Intratympanic Gentamicin Treatment." <u>Otolaryngology Head & Neck Surgery</u>. Apr 2010, Vol.142, No.4, pp 570-575.

Smith, WK *et al.* "Intratympanic Gentamicin Treatment in Meniere's Disease: Patients' Experiences and Outcomes." <u>Journal of Laryngology & Otology</u>. Jul 2006, Vol.120, pp 730-735.

Sterkers, Olivier (Editor). <u>Menière's Disease Update, 1999</u>. Kugler Publications, 2000. (Papers given at 4th International Symposium on Menière's Disease, Paris, April 1999)
- O. Michel. "Prosper Menière From Foreign Point of View." Cologne, Germany
- Takedo, T. *et al.* "Endolymphatic Hydrops Induced by Chronic Administration of
- Vasopressin." Kochi Medical School, Japan.
- Legent, François, "Prosper Menière, a Precursor Aurist." Paris, France.

Strupp, M. & Brandt, T. "Current Treatment of Vestibular, Oculomotor Disorders & Nystagmus." <u>Therapeutic Advances in Neurological Disorders</u>. Jul 2009, Vol.2, No.4, pp 223-239.

Strupp, D, Strupp M, Brandt T. "Long-term Course of Menière's Disease Revisited." <u>Acta Otolaryngolica</u>. Jun 2010, Vol.13, No.6, pp 644-651.

Tampa Bay Hearing and Balance Center Website. "Vestibular Neuritis and Labyrinthitis." Feb 16, 2012. http://www.tampabayhearing.com/contact.php

Teggi, R. *et al.* "Does Menière's Disease in the Elderly Present Some Peculiar Features?" <u>Journal of Aging Research</u>. Sep 2011, Vol.2012, Article ID 421596.

Telian, Steven. "Surgery for Vestibular Disorders." <u>Cummings Otolaryngology: Head & Neck Surgery. 5th Ed.</u>, Paul Flint, Editor; Mosley Elsevier 2010.

Telischi, FF & Luxford, MM. "Long Term Efficacy of Endolymphatic Sac Surgery for Vertigo in Menière's Disease." <u>Otolaryngology Head & Neck Surgery</u>. Jul 1993, vol 109, No.1, pp 83-87.

Thernstrom, Melanie; <u>The Pain Chronicles</u>. Farrar, Straus, & Giroux, 2010.

Thirlwall, AS & Kundu, S. "Diuretics for Meniere's Disease." <u>Cochrane Database System Review</u>. Jul 2006, Vol.3, CD003599.

Thompson, Neal. <u>Light This Candle; The Life & Times of Alan Shepard</u>. Crown Publishers, 2004.

Torok, Nicholas. "Old and New in Menière's Disease." <u>Laryngoscope</u>, Nov 1987, Vol.11, pp 1870-1877.

Tumarkin, A. "The Otolithic Catastrophe: A New Syndrome." <u>The British Medical Journal</u>. Jul 1936, Vol.1, pp 175-177.

Vestibular Disorders Association (website.) "Labyrinthitis & Vestibular Neuritis." http://vestibular.org/labyrinthitis-and-vestibular-neuritis No date. Cited 7/13.

Vestibular Disorders Association (website.) "Migraine Associated Vertigo (MAV." http://vestibular.org/migraine-associated-vertigo-mav Undated. Cited 6/13.

Vibert, D *et al.* "Diplopia From Skew Deviation in Unilateral Peripheral Vestibular Lesions." Acta Oto-laryngologica. Mar 1996, Vol.116, No.2, pp 170-176.

Walker, Mark. "Treatment of Vestibular Neuritis." Current Treatment Options in Neurology. Jan 2009, Vol.11, No.1, pp 41-45.

Wazen, Jack J. Dizzy: What You Need to Know About Managing and Treating Balance Disorders. Simon & Schuster, 2004.

WebMD (website.) "Vestibular Neuritis – Treatment Overview." Aug 2009.

http://www.webmd.com/a-to-z-guides/vestibular-neuronitis-treatment-overview

Wells, Walter A. "Dr. Prosper Ménière: A Historical Sketch." The Laryngoscope, Apr 1947, vol.7, No.4, pp 275-293. Wiley Online Library.

Wikipedia:
    Bárány, Róbert W http://en.widipedia.org/wiki/Robert_Barany
    Betahistine W http://en.wikipedia.org/wiki/betahistine
    Brainstem W http://en.wikipedia.org/wiki/Brainstem
    Caloric Reflex Test W http://en.wikipedia.org/wiki/Caloric_reflex_test
    Cerebellum W http://en.wikipedia.org/wiki/Cerebellum
    Cranial nerves W http://en.wikipedia.org/wiki/Cranial_Nerve
    Craniosacral Therapy W http://en.wikipedia.org/wiki/Craniosacral_therapy
    Dix-Hallpike Test W http://en.wikipedia.org/wiki/Dix–Hallpike_test
    Ear W http://en.wikipedia.org/wiki/Ear
    Ewald, Ernst W http://en.wikipedia.org/wiki/Ernst_Julius_Richard_Ewald
    Flourens, Marie Jean Pierre W http://en.wikipedia.org/wiki/Jean_Pierre_Flourens
    Inner ear structure W http://en.wikipedia.org/wiki/inner_ear
    MAV W http://en.wikipedia.org/wiki/Migraine-associated_vertigo
    Ménière's Disease W http://en.wikipedia.org/wiki/meniere's_disease
    Proprioception W http://en.wikipedia.org/wiki/Proprioception
    Prosper Ménière W http://en.wikipedia.org/wiki/Prosper_Ménière
    Vestibular System W http://en.wikipedia.org/wiki/Vestibular_system
    Vestibular Neuritis http://en.wikipedia.org/wiki/Vestibular_neuronitis
    Vestibulo-ocular reflex W http://en.wikipedia.org/wiki/Vestibulo-ocular_reflex

Wikipedia Images of Inner Ear. (access through Google search)

Wright, AJ. "Menière's Disease." <u>Proceedings of the Royal Society of Medicine</u>. Nov 1948, Vol.41, No.11, pp 801-805.

Yardley, Lucy. <u>Vertigo and Dizziness</u>. "Experience of Illness Series," published by Routledge, 1994.

Yazawa, Y *et al*. "Detection of Viral DNA in the Endolymphatic Sac in Menière's Disease by in situ Hybridization." <u>Journal for Oto-Rhino-Laryngology (ORL)</u>. May-June 2003, Vol.65, No.3.

CPSIA information can be obtained at www.ICGtesting.com
Printed in the USA
BVOW04s1936211015

423468BV00008B/198/P